Echinoid Palaeobiology

TITLES OF RELATED INTEREST

Physical processes of sedimentation
J. R. L. Allen

Petrology of the igneous rocks
J. T. Greensmith

Invertebrate palaeontology and evolution
E. N. K. Clarkson

A dynamic stratigraphy of the British Isles
R. Anderton, P. H. Bridges, M. R. Leeder & B. W. Sellwood

Microfossils
M. D. Brasier

Historical plant geography
P. Stott

The changing climate: responses of the natural flora and fauna
M. J. Ford

Sedimentology: process and product
M. R. Leeder

The poetry of geology
R. M. Hazen (ed.)

Sedimentary structures
J. Collinson & D. B. Thompson

Aspects of micropalaeontology
F. T. Banner & A. R. Lord (eds)

Statistical methods in geology
R. F. Cheeney

Paleopalynology
A. Traverse

Echinoid Palaeobiology

Andrew Smith

Department of Palaeontology,
British Museum (Natural History), London

London
GEORGE ALLEN & UNWIN
Boston   Sydney

George Allen & Unwin (Publishers) Ltd,
40 Museum Street, London WC1A 1LU, UK

George Allen & Unwin (Publishers) Ltd,
Park Lane, Hemel Hempstead, Herts HP2 4TE, UK

Allen & Unwin Inc.,
9 Winchester Terrace, Winchester, Mass 01890, USA

George Allen & Unwin Australia Pty Ltd,
8 Napier Street, North Sydney, NSW 2060, Australia

First published in 1984

British Library Cataloguing in Publication Data

Smith, Andrew
  Echinoid palaeobiology.
1. Sea-urchins, Fossil
I. Title
563'.95   QE783.E2
ISBN 0-04-563001-1
ISBN 0-04-563002-X

Library of Congress Cataloging in Publication Data

Smith, Andrew B.
  Echinoid palaeobiology.
(Special topics in palaeontology; 1)
Bibliography: p.
Includes index.
1. Sea-urchins, Fossil.   I.Title.   II. Series.
QE783.E2S66   1984   563'.95   83-15071
ISBN 0-04-563001-1
ISBN 0-04-563002-X (pbk.)

Set in 10 on 12 point Times by Preface Ltd, Salisbury
and printed in Great Britain by Richard Clay (The Chaucer Press) Ltd,
Bungay, Suffolk.

Foreword

Special Topics in Palaeontology was conceived as a series allowing leading specialists to present syntheses of recent advances in their fields. Most volumes will deal, therefore, with a single group of organisms, usually at class or phylum level. Others, however, may take general topics that cut across systematic groupings. In each case, the aim is to present an enduring contribution for advanced students, teachers and research workers, a reference point for current thinking and future work.

The niche we are seeking to fill is between the level of the undergraduate textbook and the specialist memoirs, monographs and journals. It is complementary to the Treatise on Invertebrate Paleontology volumes which now form such a valuable systematic data bank. In Special Topics, each volume is focused on aspects of current interest, rather than attempting an all-embracing coverage. Each is the work of a single author, or of two or three like-minded authors, presenting their own ideas and their personal evaluations of recent advances in their field. We hope these books will prove to be both informative and stimulating to new students of palaeontology and established workers alike.

In the first of this series, Andrew Smith takes a new look at the evolution of the echinoids. He has concentrated on the functional significance and adaptive success of innovations in the history of this fascinating group, and the book includes much new material as well as a reassessment of earlier ideas and controversies. Based on this evidence, a new phylogeny for the class is developed, leading to a fully revised classification of the echinoids.

Colin T. Scrutton
University of Newcastle upon Tyne
Christopher P. Hughes
University of Cambridge

Preface

In the past couple of decades there has been a major upsurge of interest in the general biology of both living and fossil echinoids. In comparison with the wealth of data on aspects such as ecology, anatomy, embryology, physiology, reproductive biology and biochemistry, available to zoologists, palaeontologists have extremely little to work from. With death, all direct evidence concerning the animal's behaviour and mode of life is lost, while the decay that follows removes all soft tissue structures. Palaeontologists must therefore apply deductive reasoning based on a knowledge of skeletal structure and function to find out how fossil groups lived. Echinoids are a particularly profitable group to work with since they have a relatively complex calcite skeleton, in which structure and function are closely related, and there is a diverse living fauna on which to study form and function. Yet, even here, the amount of biological information that can be recovered from fossil echinoids is exceedingly limited in comparison with what is available in living forms. This information is, however, invaluable as it adds the dimension of time and provides the only direct evidence we have to show how echinoids have evolved.

This book has been written with two main themes in mind. First, I have tried to show how diverse biological information can be retrieved from fossil echinoids by applying a functional morphological approach to skeletal structures. Secondly, I have tried to outline the possible biological significance of evolutionary changes in order to build up a picture of how the life-style of echinoids has diversified through time. In order to achieve these aims, I have had to include a considerable amount of original research and synthesis of disseminated data and the reader will find that some of the chapters (notably Chs 4, 7, 8 & 9) contain much that is new. In particular, both the classification scheme and the ideas about how echinoids are related to other echinoderms are fairly novel, and differ from accounts that will be found in the Treatise of Invertebrate Paleontology (Durham *et al*. 1966) and in current textbooks.

In a book of this size I have not tried to substantiate every assertion with reference to published authorities, simply to keep the bibliography within bounds. Furthermore, in some places my generalisations may have led to some oversimplification. None the less, I hope that this book will be of interest both to the palaeontologist faced with having to decide how a particular species may have lived, and to the zoologist who wishes to discover how particular biological strategies have evolved. If this book can stimulate interest in the many problems that remain unsolved then it will have served its purpose admirably.

It now simply remains for me to express my sincere thanks to all those people who have helped, both directly and indirectly, to bring about this book. In particular I should like to mention Euan Clarkson, for first kindling my interest in sea-urchins, David Nichols for being such a superb guide through the early and uncertain stages of research, and Dick Jefferies, whose astute observations and considered approach to phylogeny have had a great influence on me. For the past four years I have had the good fortune to work with Chris Paul and our countless discussions have done much to improve my understanding of echinoderms. I am also indebted to Porter Kier, whose publications have been a great inspiration to me and to all echinoid workers, and who has been helpful and enthusiastic in the preparation of this book. Last, but by no means least, I should never have written this book without the active support and enthusiasm of my wife, who not only read the manuscript and did much to remove my grammatical idiosyncracies but who also has the knack of finding better fossil sea-urchins than I ever can!

Andrew B. Smith
April 1983

Contents

1 Introduction

Today echinoids are a successful and diverse group, adapted for a variety of different life-styles and able to live in most marine habitats. Although the group is now probably past its zenith and on the decline following its post-Palaeozoic adaptive radiation, it is still an important group with just over 900 species extant. Because echinoids have a calcite endoskeleton their fossil record is good, though this has been the case only since the early Mesozoic when echinoids evolved a rigid test. Estimates by Lambert and Thiery (1909–25), Kier (1965) and Kier and Lawson (1978) of the number of valid species as of 1970 total 124 Palaeozoic species, 3672 Mesozoic species and 3250 Caienozoic species.

The echinoid skeleton is structurally complex and there is a close relationship between structure and function at all levels. The ecology of most living groups of echinoids is now fairly well understood and detailed comparison between living and fossil species can be very revealing. This approach has been successfully applied in the interpretation of fossil lineages. For example, the evolution of irregular echinoids is one of the best documented examples of adaptive radiation in any invertebrate group.

This book attempts to show how echinoids have adapted and evolved through time. In particular, I have tried to show how, through functional morphology, interpretations can be made about many aspects of a fossil's palaeobiology. Although much detailed information on echinoid morphology is introduced as and where necessary, no systematic or exhaustive treatment is attempted and those wishing for further information should consult Hyman (1955) and Durham *et al*. (1966).

1.1 General morphology

1.1.1 Skeleton

In a typical post-Palaeozoic regular echinoid such as *Echinotiara* (Fig. 1.1), the calcite skeleton or **test** is made up of many hundreds of plates arranged pentamerally. Each plate bears an assortment of **spines** and tiny pincer-like appendages termed **pedicellariae** both of which articulate upon **tubercles**. The **corona** forms the greater part of the test. In all post-Palaeozoic echinoids there are 20 columns of plates in the corona, arranged in pairs. These form the **ambulacra** (biserial columns of plates each perforated by one or more ambulacral pores) and the **interambulacra** (biserial columns of plates that alternate with the ambulacra). Plates of the corona usually interlock and are sutured together to form a rigid structure, though this is easily fragmented once the soft tissue has decayed.

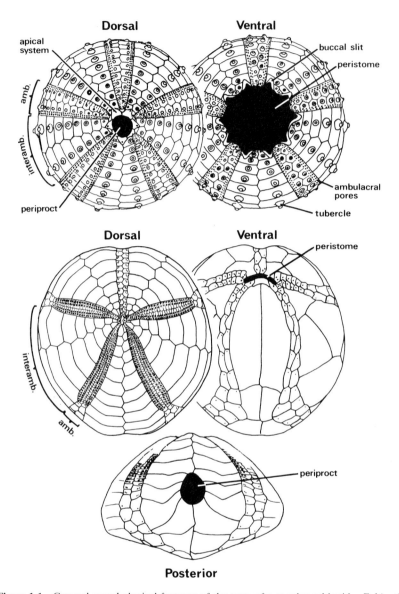

Figure 1.1   General morphological features of the tests of a regular echinoid – *Echinotiara* (upper) and an irregular echinoid – *Linthia* (lower).

   The **apical system** is situated aborally and is composed of a ring of five **genital plates**, each with a large genital pore through which gametes are released, and five so-called **ocular plates**, which are really terminal ambulacral plates and have a small pore through which the end of the radial water vessel protrudes (Fig. 1.2). The apical system encloses the **periproct** which, in life, is covered by a flexible plated membrane through

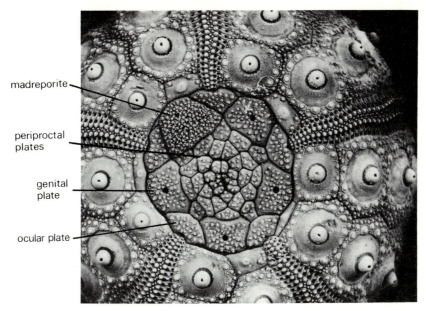

madreporite

periproctal
plates

genital
plate

ocular plate

Figure 1.2 Apical system of the Recent regular echinoid *Eucidaris metularia* (photograph courtesy of Porter Kier; approximately × 3, not in standard orientation).

which the anus opens. In some echinoids there are one or more **suranal plates** bound to the apical system.

On the oral surface there is a large **peristome** which contains the mouth and, like the periproct, is largely covered by a flexible plated membrane. The edge of the peristome may have 10 interambulacral notches, the **buccal slits**, situated immediately adjacent to the ambulacra. Within the test lies a complex dental apparatus or **lantern** (Fig. 1.3). This includes five teeth which protrude through the mouth and are used for cutting or rasping. The lantern is moved by a series of muscles that attach to internal processes developed from the inner edge of the corona. These processes form the **perignathic girdle**. All skeletal elements are composed of a three-dimensional meshwork of calcite termed **stereom**.

The plating arrangement in post-Palaeozoic echinoids is quite different from that in Palaeozoic echinoids where the number of columns of plates in ambulacra and interambulacra is quite variable (Fig. 1.6). The test has also undergone considerable modification during the evolution of irregular echinoids. In irregular echinoids such as *Linthia* (Fig. 1.1) the test has a bilateral symmetry superimposed upon the original pentamery. The periproct is no longer within the apical system but lies posteriorly and, in some groups, is even positioned on the oral surface. Spines and tubercles are usually very much smaller and more numerous than in regular echinoids and may be specialised for specific functions. Some irregular echinoids have completely lost their dental apparatus. As will be shown later, these

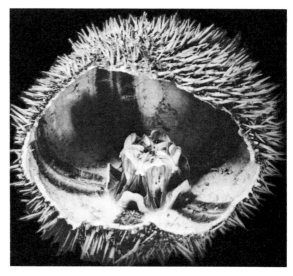

Figure 1.3 Test of *Echinus esculentus* broken open to reveal the lantern apparatus and its muscles that attach on to the perignathic girdle (× 0.66).

changes reflect the different life-styles adopted by regular and irregular echinoids. Whereas regular echinoids live epifaunally and browse or graze on algae and encrusting organisms, irregular echinoids have adopted an infaunal mode of life and are microphagous sediment eaters.

1.1.2   Soft tissue

All skeletal elements, except for the tips of the teeth and the shafts of some spines, are covered by epithelium and so form a true endoskeleton. The interior of the test is largely fluid filled and contains the various soft tissue organs (Fig. 1.4). Regular echinoids have five sac-like **gonads**, each opening to the exterior through one of the genital pores. The gonads are interradial in position and are suspended from the test by strands of connective tissue. Some irregular echinoids have five gonads but most have two, three or four.

The **digestive system** consists of a pharynx, oesophagus, large and small intestine and rectum; there is no true stomach. The pharynx lies within the dental apparatus and is absent in those groups lacking a lantern. The oesophagus is fairly long and leads directly into the intestine, which is arranged in a double U-shaped loop. The short rectum leads to the anus, situated in the periproct.

The **water vascular system** is a hydraulic system of fluid-filled tubes and sacks that is adapted for many purposes in echinoids. There is a **central ring vessel** encircling the oesophagus just above the lantern. From this an axial tube, the **stone canal**, leads directly to the **madreporite**, a modified genital plate that is perforated like a pepper pot and provides the only opening to

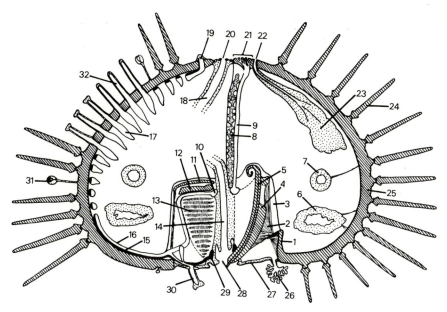

Figure 1.4 Diagrammatic cross section through a regular echinoid to show major anatomical features.

Key

1	perignathic girdle	11	compass	22	gonopore
2	lantern protractor/retractor	12	rotula	23	gonad
	muscles	13	interpyramidal muscle	24	spine
3	peripharyngeal coelom	14	oesophagus	25	corona
4	hemi-pyramid	15	radial nerve	26	compensation sac
5	tooth	16	radial water vessel	27	peristomial membrane
6	large intestine	17	ampulla	28	pharynx
7	small intestine	18	rectum	29	nerve ring
8	axial vessel	19	ocular pore	30	buccal tube foot
9	stone canal	20	anus	31	pedicellaria
10	circum-oral ring	21	madreporite	32	tube foot

the water vascular system. Five **radial water vessels** also branch from the central ring vessel and each runs up the middle of an ambulacrum immediately internal to the skeletal plates. Side branches lead from the radial water vessel to each **tube foot** and its **ampulla**. Tube feet are external organs that are connected to their internal ampullae and to the rest of the water vascular system by a single or double **ambulacral pore** that perforates ambulacral plates. The ampullae are internal fluid reservoirs that enable each tube foot to function independently of other tube feet. The radial water vessel terminates at the ocular plate where it forms a small sensory knob passing through the ocular pore.

The **nervous system** is remarkably simple and lies closely associated with the water vascular system. A nerve ring surrounds the oesophagus and gives rise to five radial nerves, each situated perradially between the radial water vessel and the inner surface of the corona. Branches from the radial nerve run to each ambulacral pore where they pass through to the exterior

of the test. At the exterior the nerve divides, part passing up the tube foot
to innervate the tip, and part spreading out beneath the epidermis of the
plate to form a sensory nerve plexus and innervate the spines and pedicellariae. Echinoids with a dental apparatus have five enlarged centres on the
central nerve ring which send nerves to the muscles of the lantern.

1.2 Classification and geological history

The higher-level classification of echinoids is based on a variety of stable
characteristics, the most important of which are lantern structure, peristomial plating and the arrangement of ambulacral and interambulacral
plates. A phylogenetic classification of echinoids is adopted here (Fig. 1.5)
and is dealt with more fully in Chapter 7.

We know only about 120 species of Palaeozoic echinoids and so the
early evolution of this group is still poorly understood. Until recently there
was a fairly clear-cut distinction between Echinoidea and other classes of
echinoderm, in that echinoids were thought to be unique in possessing a
lantern. However, the discovery of an echinoid type of lantern in Ophiocis-
tioidea (Haude & Langenstrassen 1976) and the realisation that the jaw
apparatus in Ordovician echinoids is comparable with ophiuroid jaws
has meant that Echinoidea are not so distinct as was once thought. The
relationship of echinoids to other echinoderm groups is the subject of
Chapter 9.

Echinoids first appear in the Upper Ordovician and so are one of the
youngest classes of echinoderms to have evolved. (Some workers believe
that bothriocidarids should be classed as echinoids (viz. Philip 1965,
Durham 1966) extending their range back into the Middle Ordovician,
but, for reasons outlined in Chapter 9, I prefer to exclude bothriocidarids
from the Class Echinoidea.) Although they were never particularly successful during the Palaeozoic, they rapidly diversified in the Mesozoic and
Tertiary. During their evolution, echinoids have evolved in a variety of
different ways and display quite large morphological variation (Fig. 1.6).

The earliest echinoids are **lepidocentrids** which, as currently defined, are
a miscellaneous group of all those echinoids whose only common feature is
that they lack any of the more advanced characteristics of later groups.
They include the ancestors to several later lineages. Lepidocentrids have
biserial ambulacra and a multiserial arrangement of imbricate interambulacral plates. A slightly younger group, the **echinocystitids**, have four or
more ambulacral columns and a very distinctive type of tooth structure. In
the **proterocidarids** and **lepidesthids** the number of columns of plates in
each ambulacrum multiplied considerably, and in proterocidarids the
ambulacra became greatly expanded adorally. The lepidocentrids also gave
rise to the **archaeocidarids**, a group with biserial ambulacra and large
interambulacral plates each bearing a single large tubercle. The

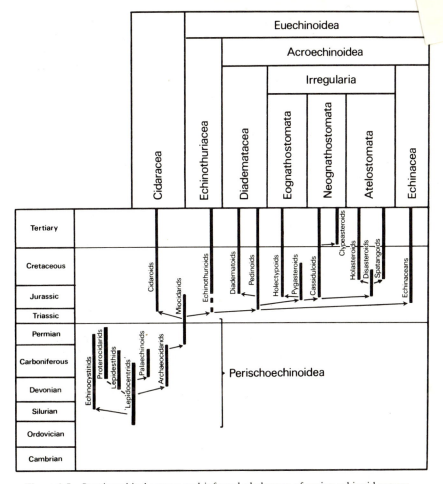

Figure 1.5 Stratigraphical ranges and inferred phylogeny of major echinoid groups.

miocidarids, an offshoot from the archaeocidarids, differ from all other Palaeozoic echinoids in having just two columns of plates in each interambulacrum. The final Palaeozoic echinoid group, the **palaechinids**, are easily identified by their greatly thickened plates that are tesselate rather than imbricate. This group is confined to the Lower Carboniferous.

Towards the end of the Palaeozoic, echinoids were clearly on the decline (see p. 149). As far as we know only one genus, *Miocidaris*, managed to survive the Permo-Triassic life crisis and saved the group from extinction. Having struggled through the crisis, however, echinoids underwent a spectacular adaptive radiation. From miocidarids the entire spectrum of post-Palaeozoic echinoids evolved. Unlike Palaeozoic echinoids, most post-Palaeozoic groups evolved a rigid test and so we know a great deal about their evolution.

During the Triassic miocidarids gave rise to two very different types of

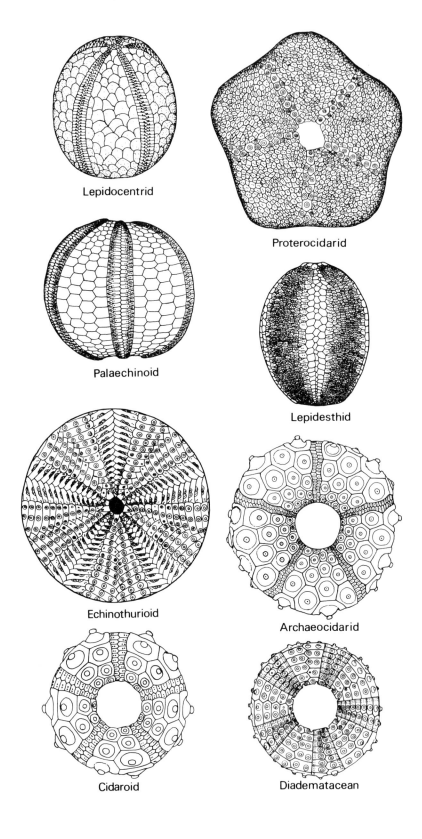

Lepidocentrid

Proterocidarid

Palaechinoid

Lepidesthid

Echinothurioid

Archaeocidarid

Cidaroid

Diadematacean

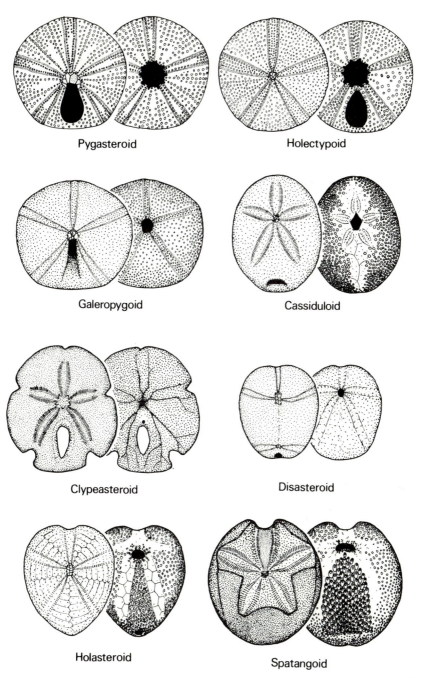

Pygasteroid

Holectypoid

Galeropygoid

Cassiduloid

Clypeasteroid

Disasteroid

Holasteroid

Spatangoid

Figure 1.6 Types of regular and irregular echinoids showing the range in morphology within the class (no scale).

echinoid, namely **cidaroids** and **euechinoids**. Cidaroids have a rigid corona of great beauty. The ambulacra are narrow and sinuous and composed throughout of simple ambulacral plates. Interambulacral plates each have a single large tubercle and spine. Cidaroids are distinguished from euechinoids by the structure of their lantern and perignathic girdle (see Ch. 7).

Amongst euechinoids, **echinothurioids** are the most primitive and today are found almost exclusively in the deep-sea environment. They have an imbricate-plated test and, as in earlier groups, a series of ambulacral plates extend over the peristome. All other regular euechinoids have more advanced lanterns than echinothurioids and have only ten ambulacral plates on the peristome. The **Acroechinoidea** comprises all euechinoids except for echinothurioids and today includes approximately three-quarters of all extant species. Acroechinoids are divided into three major groups, the **Diadematacea**, **Echinacea** and **Irregularia** on the basis of tooth structure. Diademataceans retain the primitive type of tooth which is grooved in cross section, whereas echinaceans have teeth that are keeled, and Irregularia either have teeth that are diamond- or wedge-shaped in cross section or have secondarily lost their lantern.

Irregular echinoids evolved in the early Jurassic and diversified rapidly so that they now constitute 47% of all living species. Although the periproct lies well outside the apical system in all present-day irregular echinoids, this has not always been so and the early members of this group still have their periproct fully within the apical system. The **pygasteroids** and **holectypoids**, which together constitute the group **Eognathostomata**, at first retained a functioning lantern throughout life and have relatively simple unspecialised spines and tube feet. Other irregular echinoids evolved from the **galeropygoids**, a group that arose from stem pygasteroids before the periproct had migrated out of the apical system. The **atelostomates** are a group of typically heart-shaped echinoids that have lost all trace of a lantern: it includes the **spatangoids**, which have a compact apical system, and the **holasteroids**, which have a more elongate arrangement of apical plates. Both spatangoids and holasteroids arose in the early Cretaceous from the **disasteroids**, a group with a split apical system. **Neognathostomates** include both the **cassiduloids** and the **clypeasteroids**. Cassiduloids resemble galeropygoids but have petaloid ambulacra. Except for one small group, cassiduloids have a lantern only when they are juveniles. The clypeasteroids first appeared in the Palaeocene and so are the most recent group to have evolved. Sand-dollars belong to this group, whose most distinctive feature is the presence of enormous numbers of tiny tube feet and ambulacral pores on ambulacral plates outside the petaloid areas.

2 Taphonomy

Although it may seem strange to start this review of echinoid palaeobiology with a discussion of death, decomposition and diagenesis, it is important to understand the changes that can take place after death when interpreting fossil material. For example, the ease with which the skeleton disintegrates following death affects the preservation potential. This varies from group to group, and the preservation potential of echinoids in general has changed through time. The way in which a fossil echinoid is preserved may also be a useful guide as to how it met its death.

2.1 Death

There are many ways in which echinoids may die but the principal causes of death are predation, storm action, exposure to high temperatures and senescence.

2.1.1 Predation

Post-larval echinoids are preyed upon by a wide range of animals including other echinoids, starfish, gastropods, decapod crustaceans, octopuses, fish, birds, otter and man (Moore 1966). Predation will rarely decimate a population and under normal conditions accounts for only a small proportion of deaths. The regular echinoid *Strongylocentrotus droebachiensis*, for example, has many predators yet the abundance of this urchin is not obviously affected by predation (Himmelman & Steele 1971).

Fish are arguably the most important predators of echinoids today. Some 27 species of fish from the Caribbean and nine species from the Red Sea have been observed feeding on regular echinoids (Randall 1967, Fricke 1971). Most of these fish devour echinoids whole. Attacks on juvenile *Diadema* and the short-spined *Echinometra* are usually successful but few fish can tackle adult diadematids which have long, needle-sharp spines, or the stout-spined *Heterocentrotus*. To avoid the echinoid's defensive array of spines, fish usually try to overturn the sea-urchin and attack the peristome where protection is minimal. Some fish clip off the spines before swallowing the test and leave behind a strewn pile of broken spines. Species of Labridae will even pick up large urchins in their mouths and break them into manageable pieces by pounding them against rocks.

Crabs and lobsters can also be important predators on echinoids. Both use a similar method of attack. Regular echinoids are prised off the substratum and turned over. A hole is then made through the oral surface using the claws, either through the peristomial membrane or through the corona. Finally, the claws are used as pincers and the test cracked open and broken into a number of irregular fragments. Crabs will also attack sand-dollars, using their claws to crack open the test from the margin inwards. Not all attacks are successful and, in a population of the Recent sand-dollar *Dendraster*, some 5% had abnormally configured tests attributed to unsuc-cessful crab attacks (Birkeland & Chia 1971). Similar abnormalities are not unusual in fossil sand-dollars and over 50% of the population of the Miocene sand-dollar *Monophoraster* from one locality in Patagonia examined by Zinsmeister (1980) showed evidence of predation (Fig. 2.1).

Sand-dollars and regular echinoids rarely occur in the same habitat but where they do, regular echinoids will attack sand-dollars (Himmelman & Steele 1971). *Strongylocentrotus* starts its attack on the sand-dollar *Echinarachnius* by eating the marginal spines. It then moves adapically and scrapes an irregular hole through to the interior which is later widened. Starving regular echinoids in natural populations have even been observed to cannibalise one another. In such attacks the predator clips the spines from a small area and makes an irregular hole through the test, usually in the apical region, which is enlarged by chewing radial tracts (Fig. 2.2).

Starfish attack both regular and irregular echinoids. They swallow small echinoids whole and wrap themselves around the larger ones. As the soft tissue is digested the spines fall off and the starfish eventually moves on leaving a clean, intact test free of spines.

Large predatory gastropods have been observed feeding on both infaunal and epifaunal echinoids (Moore 1956). Once it has located its prey, the gastropod clasps the echinoid in its foot and a small aboral or anterior area is cleaned of spines. Then a neat hole is drilled through the test and the soft parts digested, leaving an otherwise intact test (Fig. 2.3). A similar method of attack is used by octopuses.

Figure 2.1 A specimen of the Miocene sand-dollar *Monophoraster darwini* (dorsal surface) with a rather irregular outline due to predation (× 1). There are three separate damaged regions (arrowed) that have been repaired (photograph courtesy of W. J. Zinsmeister and the Geological Society of America).

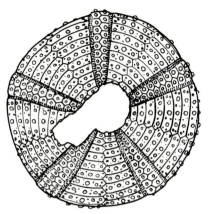

Figure 2.2 Outline drawing of a test of the echinacean *Strongylocentrotus droebachiensis* (Recent) that has been cannibalised by another individual (× 1) (taken from a photograph in Himmelman & Steele 1971).

Evidence of gastropod predation is rare in fossil echinoids and the earliest undoubted predation boreholes that I have seen have been in Upper Cretaceous (Coniacian) *Micraster turonensis* from Sarthe in France, where occasional specimens may have a single neatly drilled hole dorsally. Kier (1981) has reported possible drill holes in an Upper Albian (Lower Cretaceous) spatangoid, *Hemiaster*.

Parasitic gastropods may also attack sea-urchins, but they do not generally kill their hosts. The holes these parasites produce are quite different from those of predators since they tend to be more irregular and characteristically have abnormal stereom deposition around the wound (proving that the parasite was attached to the living animal for some time). Kier (1981) has also described parasitised heart-urchins from the Lower Cretaceous.

Echinoids that live intertidally or immediately subtidally may be

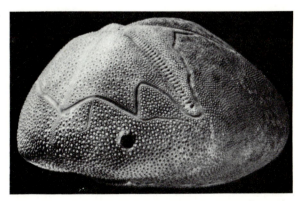

Figure 2.3 Lateral view of the spatangoid *Meoma ventricosa* (Recent) with a circular borehole made by a predatory gastropod at the ambitus of interambulacrum 2 (× 0.5).

Figure 2.4 Part of the test of the echinacean *Echinus esculentus* (Recent) dropped and shattered on the foreshore by seagulls (approximately × 1). Notice how fractures cross plates and follow lines of ambulacral pores rather than breaking along sutures.

attacked by birds. Different birds use different methods of attack. Eider ducks will swallow relatively small echinoids whole. Turnstones overturn regular echinoids and peck through the peristomial membrane (Glynn 1968) and the same method is used by herring gulls attacking *Strongylocentrotus*. In Britain, the large regular echinoid *Echinus* is picked up by herring gulls and dropped from a height on to rocky foreshore pavements in order to shatter the test (Fig. 2.4).

Finally, the otters of California are renowned for the way in which they feed on regular echinoids (Hall & Schaller 1964). The otters dive and collect one or two echinoids together with a stone from the bottom. On surfacing they roll on to their backs and proceed to break open the urchins by pounding them (against their chests!) with the stones.

2.1.2 Storms

Severe storms can have a catastrophic effect upon shallow-water echinoid populations, sometimes causing mass mortality. During storms, large quantities of sediment are shifted and may bury animals to depths from which they cannot escape. A covering of just 30 cm will entomb both regular and irregular echinoids (Schäfer 1971). Storms may also wash out and transport echinoids to other environments. Storms can wash large numbers of shallow-water echinoids on to beaches where they are quickly pulverised. Others may be washed out to sea and dropped below the wave base (Parks 1973). The regular echinoids that live epifaunally on shallow, rocky substrata usually have a powerful enough grip not to be dislodged during storms, but they face the danger of being crushed *in situ* by transported debris.

2.1.3   High temperatures and desiccation

Another common cause of mass mortality in shallow-water populations is prolonged exposure to elevated temperatures (Glynn 1968, Hendler 1977). This is likely to occur on reef flats when low spring tides coincide with midday in spring or summer. On these occasions there are reports of up to 86% of the population being killed. Echinoids living in shallow lagoons and pools are as much at risk as stranded echinoids, and infaunal echinoids are no more protected than epifaunal echinoids. Exposure to air for just two to three hours will kill most echinoids, including infaunal heart-urchins. Echinoids in crevices out of the direct heat will last a little longer (three to four hours). As an echinoid becomes desiccated, air enters the test so that, when the tide returns, the test starts to float. Even exposure for a short period can cause sufficient desiccation to make the test buoyant and will eventually lead to death. Shallow, enclosed pools and lagoons may also become overwarm. Echinoids in water at a temperature of 35°C will survive just three hours or so. Death does not occur through desiccation and so the tests do not float.

Salinity changes that can occur in shallow pools during heavy rain storms rarely result in death. Echinoids can survive two or three hours in 50% sea water without ill effect (Giese & Farmanfarmaian 1963).

2.1.4   Senescence

Old age seems to be the principal cause of death in the regular echinoids *Paracentrotus lividus* (Crapp & Willis 1975) and *Strongylocentrotus intermedius* (Kawamura 1973) and in the sand-dollar *Dendraster excentricus* (Birkeland & Chia 1971). Echinoids generally have a life span of 1–15 years (see Table 3.1, p. 88) and senescence is probably a fairly common cause of death in many other species. In short-lived and rapidly growing opportunistic species there may be a mass mortality following spawning.

2.2   Decomposition

The way in which the skeleton of a regular echinoid disintegrates following death has been recorded by Schäfer (1971). In the final stages of death the spines and pedicellariae become moribund, and droop to lie flattened against the test. Once decay has set in, first the pedicellariae and then the spines become detached and lie scattered around. Within a week most of the spines have fallen off and within two weeks the peristomial and periproctal membranes have disintegrated. The lantern has started by now to fall into pieces and is lost through the peristome. In some species the plates of the apical system may also fall in. By this time the connective tissue that binds the plates of the corona together is sufficiently decayed that any relatively minor disturbance will cause the corona to fragment along sutures. The activity of scavengers greatly speeds up the process of disintegration.

Irregular echinoids have smaller spines bound with relatively little organic material and their spines are lost within a matter of hours after death. Once the connective tissue that binds the plates together has decayed, the corona is held together because the stereom of adjacent plates interdigitates slightly across sutures (Fig. 2.5). Some groups, such as diadematoids and cidarids, have almost no stereom interlocking across sutures and will collapse into a pile of plates even when undisturbed. Other groups have better interlocking plates and will remain intact so long as they are not disturbed too much. In clypeasteroids, interlocking is extensive and the plates remain firmly bound together even after all soft tissue has disappeared. The clean test is fairly robust and can survive transportation. This

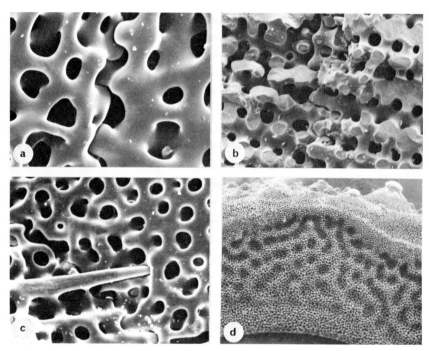

Figure 2.5 Scanning electron micrographs of stereom interlocking at plate junctions. (a) The cidarid *Cidaris cidaris* (× 475) where stereom interdigitation between plates is limited. (b) The echinacean *Colobocentrotus atratus* (× 300) with much more extensive stereom interpenetration at plate junctions. (c) The sand-dollar *Mellita quinquiesperforata* (× 300) where large interpenetrating thorns bind adjacent plates together. (d) The lateral face of a plate of the echinacean *Temnopleurus hardwicki* (× 30) showing the peg and socket development of stereom that ensures that adjacent plates interlock securely.

explains why they are the only group of echinoids that can be preserved as accumulate coquinas.

2.3 Interpreting preservation

The sutures are greatly strengthened by the sutural connective tissue binding plates together (see p. 31) so that any breakage of the test that takes place while the animal is alive crosses through plates rather than following sutures. At death the sutural tissue starts to decay and any buffeting will break the test along suture lines. Hence there should be a correlation between the state of preservation and the manner of death. Fossil echinoids are usually preserved in one of the following ways.

(a) *Whole test complete with spines, pedicellariae and lantern* (Fig. 2.6a). Death has occurred as a result of rapid sedimentation burying the echinoid to depths from which it cannot escape. Spines are held in position by the enveloping sediment even after all the supportive soft tissue has decayed. Should the sediment be reworked and the test exhumed then it would simply disintegrate into its constituent plates. Most Lower Palaeozoic echinoids are preserved in this way but such preservation becomes relatively rare in the Mesozoic and Tertiary.

(b) *Whole corona with or without apical plates, lacking periproctal and peristomial plates, spines and lantern* (Fig. 2.6c, d). Here death could have occurred from various causes (asteroid predation, mass mortality, senescence) but the test has not been buried until some time after death. Where there is little disturbance the corona, freed of spines and lantern, may rest on the sea floor for months and become encrusted before final entombment within the sediment. The extent to which stereom interlocks across sutures controls the amount of buffeting that the test can withstand, and the length of time before disintegration occurs after death. This type of preservation is relatively common in quieter areas of sediment accumulation from the Jurassic onwards.

(c) *Coronal segments separated along plate sutures, and single plates* (Fig. 2.6e). Conditions as in (b) but post-mortem disarticulation has continued further. This may be because of weaker plate interlocking (for example, cidarids disarticulate more readily than echinaceans) or due to more turbulence or scavenging.

(d) *Localised pile of dissociated plates, spines and lantern elements* (Fig. 2.6b). Death has occurred in a tranquil environment through senescence or some natural catastrophe (but not by storm action). The echinoid has disintegrated to a pile of plates as the soft tissue decayed and there has been little disturbance. Echinoids preserved in this way usually have little or no stereom interlocking between their plates.

Figure 2.6 Preservational styles. (a) *Acrosalenia pustulosa* (Middle Jurassic): the entire test is preserved with spines, pedicellariae and lantern in position (approximately × 2). (b) *Diademopsis tomesii* (Lower Jurassic) preserved as a strewn pile of plates, spines and lantern elements (× 1). (c) *Clypeus ploti* (Middle Jurassic): the test has been bored and is encrusted by serpulids and oysters (× 1). (d) *Acrosalenia lycetti* (Middle Jurassic): only the corona is preserved, spines, lantern and apical system having been lost (× 1.5). (e) *Acrosalenia lycetti* (Middle Jurassic): a segment of the corona separated along plate sutures and consisting of two ambulacral and two interambulacral columns (× 2.5). (f) *Plesiechinus ornatus* (Middle Jurassic): an irregular fragment of test broken across plate sutures (approximately × 1).

(e) *Test fragments irregularly broken across plates, with or without spines* (Fig. 2.6f). Test breakage must have occurred while the echinoid was alive and sutural connective tissue could still function in transmitting stress. Death has occurred either through predation (by crustaceans, other echinoids, birds or otters) or by pulverisation in a turbulent environment. Distinguishing between predation and crushing is difficult. Only a small proportion of a fossil population might be expected to die from predation and be preserved in this way, whereas the entire population might be decimated by storm action to produce a concentration consisting largely of fragmental debris.

Studies dealing with echinoid preservation are few, but one good example by Aslin (1968) looked in detail at the Middle Jurassic regular echinoid *Acrosalenia*. Aslin found that in one quarry specimens of *Acrosalenia* were preserved in one of two ways: (a) as a whole test with spines, apical disc and lantern, usually filled with sparry calcite; and (b) as a whole sediment-filled corona or as fragments, lacking spines, apical disc and lantern. Each was found to occur in a different type of sediment. Aslin showed that the whole tests complete with spines represented individuals that had been overwhelmed by a sudden slurry of semi-plastic sediment that buried and killed them. The other, more usual style of preservation resulted when echinoids died of natural causes and decomposed on the sea floor before becoming incorporated into the sediment.

Another very elegant study, on the preservation of the Lower Liassic regular echinoid *Diademopsis* from a small locality near Tübingen, was published by Bloos (1973). Bloos demonstrated that these echinoids had had a complex taphonomic history that included several phases of sedimentation. Using the rate at which Recent regular echinoids disintegrate, he was able to estimate the length of time between successive pulses of sedimentation by noting, on specimens that were initially only partially buried, how much further decay had proceeded on the upper surface compared with the lower surface.

In the Middle Jurassic of the mid-Cotswold Hills, England, a varied fauna of echinoids in various states of preservation can be collected from what is loosely known as the Pea Grit Series. Some 17 species of echinoid occur in this diverse carbonate sequence, all within the *Murchisonae* Zone of the Bajocian. Echinoids are confined principally to four lithofacies within the Series:

(a) *Upper Beds of the Crickley Limestone.* Oobiosparites containing irregular echinoids and interpreted as open platform oolite sands (Mudge 1978) with thin, bioturbated marl horizons containing pygasteroids and regular echinoids produced during periods of bottom stabilisation.
(b) *Crickley Oncolite.* This is a highly bioturbated oncolitic biomicrite with numerous encrusted intraclasts of local intrabasinal origin. It is

interpreted as a stable, rubbly algal gravel deposited in a tranquil environment.

(c) *Crickley Coral Bed*. A highly bioturbated biomicrite with patchy development of corals, interpreted as a lagoonal patch reef environment (Mudge 1978).

(d) *Top Beds of the Cleeve Hill Oolite*. An oobiomicrite with abundant bryozoan debris, interpreted as toeset beds of a sand shoal deposited offshore to a shallow firm bottom.

The distribution of species and their style of preservation amongst these four lithofacies is graphically illustrated in Figure 2.7. Echinoids occur as

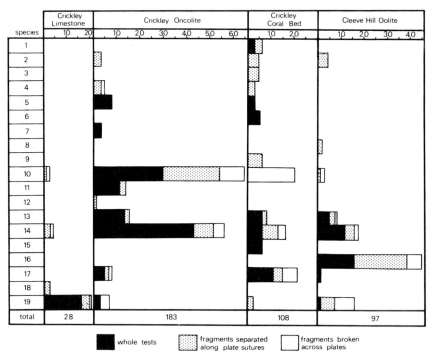

Figure 2.7 Preservational style in four lithofacies within the Pea Grit Series (*Murchisonae* zone, Bajocian, Middle Jurassic) of the mid-Cotswolds, England. Species 1–19 as follows:

Cidaroida: 1 *Rhabdocidaris fowleri* (Wright); 2 *Procidaris bouchardi* (Wright); 3 *Plegiocidaris wrighti* (Desor); 4 *Plegiocidaris* sp.

Pedinoida: 5 *Hemipedina perforata* (Wright); 6 *Hemipedina* aff. *waterhousei* (Wright); 7 *Palaeopedina bonei* (Wright); 8 *Palaeopedina bakeri* (Wright); 9 *Pseudopedina* cf. *divionensis* (Michelin).

Pygasteroida: 10 *Plesiechinus ornatus* (Buckman); 11 juvenile pygasteroids.

Order uncertain: 12 *Heterocidaris* sp.

Echinacea: 13 *Acrosalenia lycetti* (Wright); 14 *Trochotiara depressus* (Agassiz); 15 *Psephechinus deslongchampsi* (Wright); 16 *Stomechinus intermedius germinans* (Phillips); 17 *S. intermedius* (Agassiz) (juveniles).

Cassiduloida: 18 *Clypeus michelini* (Wright); 19 *Galeropygus agariciformis* (Forbes).

(a) whole coronas with or without an apical system but lacking spines and lantern, (b) corona fragments separated along plate sutures, or (c) corona fragments broken across plates. Unlike the population studied by Aslin, none of the material from the Pea Grit Series represents animals that had been entombed by a rapid influx of sediment. Indeed many of the specimens appear to have spent some time exposed on the sea floor after death and have begun to become encrusted. A small proportion of specimens in all lithofacies occurs as fragments broken across plate sutures. These may be the remains of predated individuals or may be fragments of storm-crushed individuals washed into areas of sediment deposition.

The distribution of the pygasteroid *Plesiechinus* is worthy of mention. It is clear that *Plesiechinus* lived on the algal gravels forming the Crickley Oncolite since large numbers of whole tests of all sizes are found here. *Plesiechinus* also occurs not uncommonly in the Crickley Coral Bed, but here it is represented almost entirely by test fragments fractured across plates. The most likely explanation for this is that the fragments represent individuals, possibly living at the periphery of the main population, that have been washed out in storms and crushed. The fragments were then transported leeward into the protected lagoonal environment where they were preserved.

2.4   Diagenetic changes

The changes that take place after burial are of less interest than those occuring prior to burial as they can tell us nothing about the environment in which the fossil lived. Very early on in diagenesis the echinoid skeleton is transformed from its original high-magnesium calcite to a low-magnesium calcite, but this has no visible effect on skeletal ultrastructure. The highly porous skeleton of stereom is often infilled with secondary calcite, deposited in optical continuity with the original calcite. However, where the skeleton is buried in a mud or fine silt, the pore space within plates can become filled with sediment. Should the calcite later be lost, a three-dimensional mould of the stereom is preserved in sediment.

Diagenetic fracturing is a common feature in fossil echinoids. For example, many of specimens of the cassiduloid *Clypeus ploti* from the Middle Jurassic Clypeus Grit of England have been fractured around the margin of the oral surface and the oral surface pushed inwards. This has happened because the tests, owing to their relatively tiny periproct and peristome openings, were not entirely filled with sediment and collapsed on compaction. As breakage of the test is irregular and does not follow sutures, and as encrusting organisms are equally affected, fracturing must have taken place after an early phase of cementation had strengthened the test.

2.5 Preservation potential

Today most echinoids stand a much better chance of being preserved than
any other type of echinoderm because of their rigid test. However, this has
not always been the case. Palaeozoic echinoids are relatively rare and the
great majority of Lower Palaeozoic echinoids come from just a handful of
localities where there have been unusual conditions for preservation. Many
Palaeozoic echinoids had imbricate plated tests and so dissociated rapidly
upon death. Unless buried alive, these echinoids were unlikely to be preserved except as isolated plates. The same is true of the imbricate plated
echinothurioids of today. Echinothurioids are believed to have evolved in
the late Triassic – yet they have a very poor fossil record. There are 47
species of echinothurioids alive today, yet we know of just eight fossil
species: two from the Jurassic; four from the Upper Cretaceous; one from
the Palaeocene; and one from the Miocene. The poor fossil record of
echinothurioids is also due to the fact that the group lives almost exclusively in the deep-sea environment, which is rarely represented in the
geological record. Tertiary holasteroids are also greatly under-represented
in the fossil record for the same reason (Kier 1977).

Kier (1977) noticed that regular echinoids appear to have a relatively
poor fossil record when compared to irregulars. Only 20% of known Tertiary echinoids are regular whereas, today, 53% of echinoid species are
regular. Kier attributed the poor fossil record of regular echinoids largely
to the fact that regular echinoids live epifaunally and are exposed to currents and scavengers whereas irregular echinoids live infaunally. However,
the great majority of fossil irregular echinoids are preserved without their
complement of spines and so are unlikely to have been buried alive. When
placed under stress, irregular echinoids will come up to the sediment surface where they commonly die. The crucial difference between regular and
irregular echinoids is in their life-styles. Since the Jurassic, regular
echinoids have evolved and diversified as grazers, many adopting
shallow-water habitats and living on firm or rocky substrata. In contrast,
irregular echinoids evolved and diversified as deposit feeders and came to
live in areas of unconsolidated sediment. Regular echinoids have a relatively poor fossil record because they came to live predominantly in areas
of active erosion where they stood little chance of being preserved. Irregular echinoids on the other hand inhabited areas of active sedimentation and
their fossil record is good.

The likelihood of an echinoid being preserved depends not only upon
the type of environment it inhabits but also upon how readily the test
disintegrates after death. In those echinoids with an imbricate plated test,
the plates are embedded in connective tissue and there is no stereom
interlocking between plates. As the soft tissue decays, the test simply col-
lapses into a pile of plates. Echinoids with a rigid test are more likely to be
preserved, but even here not all groups have the same preservation poten-

Table 2.1 Number of echinoid species in the Miocene and Recent (from Kier 1977). Classification as in Durham *et al.* (1966).

	Number of Miocene species	Number of Recent species
Regular orders		
Cidaroida	45	144
Echinothurioida	1	47
Diadematoida	8	48
Salenioida	3	14
Pedinoida	0	10
Temnopleuroida	71	119
Phymosomatoida	0	2
Echinoida	50	65
Arbacioida	7	25
Irregular orders		
Holectypoida	8	3
Cassiduloida	157	28
Clypeasteroida	408	129
Holasteroida	3	30
Spatangoida	348	236

tial. If you compare the number of species known from the Miocene with those alive today (Table 2.1) it is apparent that the Echinoida and Temnopleuroida have better fossil records than other groups of regular echinoid. This reflects the extent of stereom interdigitation that occurs between abutting plates and holds the test together once all soft tissue has decayed. Temnopleuroids differ from all other regular echinoids in having peg and socket structures on plate suture faces (Fig. 2.5d), and these bind the plates firmly together after death. Echinoida lack peg and socket jointing but stereom knobs interlock more extensively between plates in this group than in cidarids, diadematids, pedinids or saleniids.

Amongst irregular echinoids, clypeasteroids have undoubtedly the best preservation potential because of the deeply interpenetrating stereom pegs and rods that bind the plates together even after death (Fig. 2.5c). The holasteroids are poorly represented in the Miocene because they are found living principally in deep-sea habitats.

The preservation potential for echinoids therefore depends upon (a) the rigidity of the test and (b) the type of environment inhabited, and has changed dramatically through time, which is why we know a great deal more about the post-Palaeozoic evolution of this group than we do about its Palaeozoic history.

3 Adaptation through time

3.1 Skeletal structure and design

The echinoid skeleton is a true endoskeleton that is formed mesodermally and provides both protection and support. For the skeleton to function successfully in these roles certain constructional and mechanical conditions have to be met. These affect not only how the skeletal elements are arranged but also how individual elements are built. Many structures that have evolved are mechanically the most apt designs, and so analysis for structural efficiency often helps to explain why particular lines of evolution were followed. It is these aspects of skeletal design that will be considered here.

3.1.1 Composition and structure of the skeleton

All echinoderms have a skeleton composed of a three-dimensional meshwork termed **stereom**, and echinoids are no exception. Although each plate is formed of a single calcite crystal the stereom shows no hint of cleavage. The crystallographic *c*-axis may be horizontal, perpendicular or oblique to the outer surface of plates, a feature which may have some phylogenetic significance (Raup 1962) but has no obvious functional role. In curved spines the *c*-axis is warped so as to remain parallel to the length of the spine. The interconnecting pore space within stereom is filled with an organic matrix of connective tissue termed **stroma**. The porosity of the skeleton is variable even within single elements and ranges from almost nil to more than 60% of the total volume.

Stereom is composed of a thermodynamically metastable high magnesium calcite. The amount of magnesium incorporated into the calcite is quite variable, but in most elements magnesium content ranges from about 5 to 15% of the total weight of calcite (Weber 1969). Factors such as the age and growth rate of an individual are more important than environmental factors in determining the level of magnesium incorporated into the skeleton. Magnesium levels are known to increase directly as growth rate increases (Davies *et al*. 1972, Weber 1973). There is one area strikingly enriched in magnesium: magnesium carbonate attains levels of about 40 mol % in a small part of the tooth known as the stone zone. This zone of protodolomite is very narrow (Fig. 3.1) and cannot reflect a difference in growth rate. The preferential incorporation of high levels of magnesium into the stone zone makes it distinctly stronger than any other region of the skeleton. Since the stone zone forms the working tip of the tooth its strength is crucial and its selective advantage obvious.

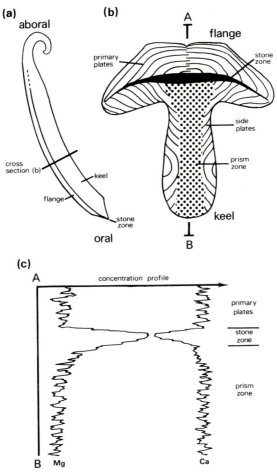

Figure 3.1 Levels of calcium and magnesium in the echinoid tooth: (a) side view of a keeled tooth; (b) diagrammatic cross section of the tooth to show the different structural areas; (c) qualitative concentration profiles for calcium and magnesium taken approximately along section A–B (from Märkel *et al.* 1977, courtesy of Gustav Fischer Verlag).

Stereom also contains other elements in small quantities; iron and strontium at less than 1% weight and manganese, aluminium and silicon at less than 100 p.p.m. (Weber 1969). Variations in their concentration are again controlled by physiological rather than environmental factors.

Not all skeletal elements are composed entirely of a single calcite crystal. Polycrystalline calcite has been found in four areas of the test: (a) at the articulation surfaces of the tubercle mamelon and the base of the spine; (b) at the articulation surfaces between rotulae and epiphyses in the lantern; (c) forming the outer cortical layer of some spines; (d) forming the secondary stereom that binds elements of the tooth together. Märkel *et al.* (1971) have shown that this polycrystalline calcite is a primary structure and not a secondary feature produced by fragmentation and repair of monocrystalline stereom. They have also shown that areas of polycrystalline calcite are harder than monocrystalline areas, which explains why they are developed at bearing surfaces and other areas where strength is required.

3.1.2   Advantages of stereom

Echinoderms are unique in having a skeleton composed of stereom and it is worth considering why such an arrangement should have evolved. There are a number of possible advantages in constructing the skeleton in this way. Stereom is highly economical in calcium carbonate which allows the animal to grow rapidly. It also produces a lightweight structure which is an advantage for mobile animals. It is particularly important that spines, pedicellariae and lantern elements are strong yet light in order to minimise the energy expended in moving them. The porous structure of stereom is also ideally suited for the insertion of fibrous tissue such as muscle and collagen which can be looped around the trabeculae for anchorage. Currey and Nichols (1967) pointed out yet another advantage: any fractures that develop in trabeculae cannot easily be propagated since they immediately come up against a stroma-filled gap. The spread of hairline fractures caused by failure under stress is inhibited because of the way in which stereom is designed, and the plate is less likely to crack than one made of solid calcite. In structure, echinoid stereom is analogous to man-made composite materials in which 'whisker' high purity single crystals are embedded in a continuous matrix. This arrangement imparts the theoretical maximum strength of the stronger material to the composite, while inhibiting crack propagation and fracturing (Weber et al. 1969). One final advantage of stereom as a building material is that the fenestrate structure has high shock resistance capabilities. Localised impact loading of the skeleton can be transmitted rapidly over a large area and the energy of impact can be dissipated in small-scale fracturing.

The way in which stereom is arranged in three dimensions has great functional significance. So far nine different arrangements of stereom have been identified in echinoids (Fig. 3.2). Precisely which of these types of stereom develops appears to be controlled by two principal factors: (a) the direction and rate of plate growth; and (b) the nature of the associated soft tissue. This is extremely useful since stereom structure can be used to reconstruct soft tissue anatomy in fossil echinoids and to interpret their growth strategies.

Galleried stereom is developed wherever collagen has to attach on to an actively growing surface. Collagen fibres are arranged into bundles which penetrate deep into the stereom along the galleries of aligned pores (Fig. 3.3a, c). The bosses of tubercles and the suture faces of plates are usually composed of galleried stereom.

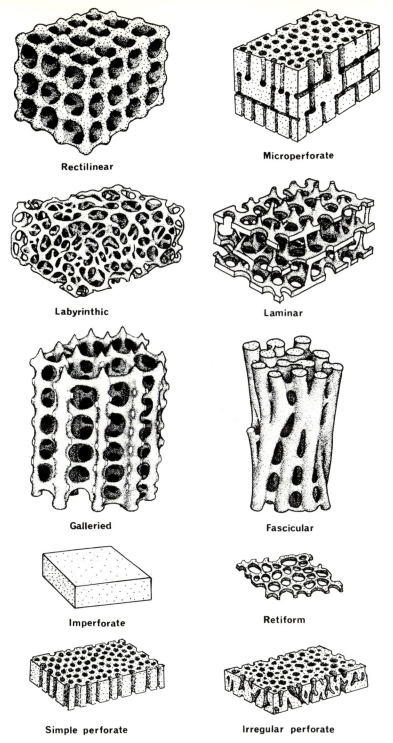

Rectilinear

Microperforate

Labyrinthic

Laminar

Galleried

Fascicular

Imperforate

Retiform

Simple perforate

Irregular perforate

Figure 3.2 Block diagrams of the different three-dimensional arrangements of stereom that occur in echinoid plates (from Smith 1980c, courtesy of the Palaeontological Association).

Muscle is attached to stereom in rather a different way to collagen. Individual muscle fibres are anchored to a thin collagenous ligament which is wrapped around the surface layer of stereom. Muscle fibres generally attach to surfaces where there is little or no outward growth taking place. The stereom where muscle attaches is no more than a superficial layer of fine retiform or labyrinthic stereom on the outer surface of the plate (Fig. 3.3b, d).

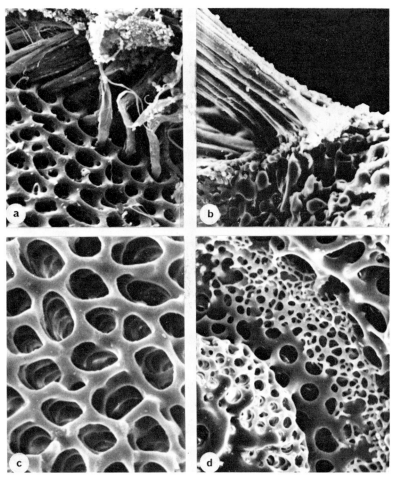

Figure 3.3 Different types of tissue attach to the skeleton in different ways to produce distinctive stereom arrangements. (a) Bundles of collagen fibres penetrate deep into the skeleton (tubercle boss of *Psammechinus miliaris*, × 750). (b) Muscle fibres are anchored to a superficial platform of stereom (tubercle areole, × 750). (c) Where collagen inserts, the stereom is galleried and shows good pore alignment perpendicular to the surface (tubercle boss of *Tripneustes gratilla*, × 500). (d) Where muscle attaches there is a superficial layer of retiform or labyrinthic stereom. Part of the tubercle of *Encope michelini* showing galleried stereom of the boss (lower left), surrounding fine retiform layer of the areole, and coarse labyrinthic stereom of the main body of the plate, × 450. All figures are scanning electron micrographs. (a and c from Smith 1980c, courtesy of the Palaeontological Association.)

A coarse labyrinthic stereom is extensively developed as a 'filler', increasing the thickness of the plate centrally to match the thickening that takes place peripherally as the plate grows in size. This is much coarser than the labyrinthic stereom associated with muscle attachment. Microperforate stereom and laminar stereom are also deposited as 'fillers' and consist of clearly defined layers of stereom arranged parallel to the surface of the plate. These arrangements may give rather more resistance to impact loading than labyrinthic stereom, an important factor in thin plates.

Fascicular stereom is found where stereom has had to grow rapidly in one direction, as for example in forming the internal pillars of the clypeasteroid test or the core to large tubercles. Imperforate stereom is usually polycrystalline and is the typical stereom of bearing surfaces at articulation points. However, it is also found elsewhere and forms the monocrystalline plates that make up the tooth. Perforate stereom is developed as an outer 'crust' to surfaces that have temporarily or permanently ceased to grow. Perforate stereom provides a stronger and more resistant surface providing protection against abrasion. In addition, the external surface of a plate is normally covered in tiny pegs or spikes of stereom. These are believed to break the shock of localised impact loading since the spikes will fracture before the stereom of the outer plate layer is damaged.

Where poor preservation has destroyed the internal stereom structure of the test, it is often possible to use a graph of average trabecular thickness plotted against average pore diameter of stereom measured at the plate surface to distinguish areas associated with different tissues. Stereom structure provides vital clues about soft tissue anatomy in fossil echinoids and allows areas of muscle and collagen attachment to be distinguished. Stereom is particularly useful when inferring tube foot anatomy from ambulacral pore morphology or spine function from tubercle structure.

3.1.3   Plate structure

In the great majority of present day echinoids, coronal plates are firmly sutured together with collagen fibres to form a rigid test. In a cross section, such plates have two or, more usually, three distinct layers (Fig. 3.4). There is a small central core of labyrinthic stereom which is the first part of the plate to form. Much of the rest of the plate is composed of galleried stereom which was laid down as the plate enlarged by peripheral accretion. Galleried stereom is formed here because of the presence of penetrating connective tissue fibres which run between plates, 'sewing' them together. Collagen fibre orientation determines the direction of pore alignment in the galleried stereom and is perpendicular to each sutural face. As the plate increases in size it also increases its thickness peripherally. In order to maintain a uniform plate thickness, the central region of the plate is also thickened by the deposition of an internal 'filler' (usually a labyrinthic stereom). In many species a dense layer of stereom forms an outer surface

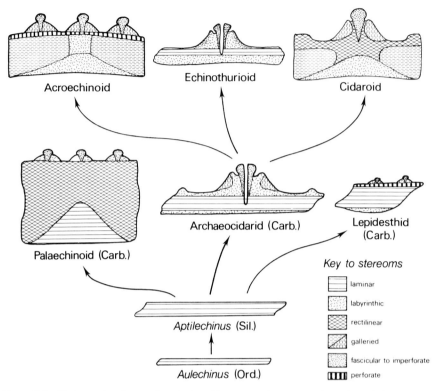

Figure 3.4   Plate construction and evolution. Diagrammatic cross sections through interambulacral plates (external surface uppermost) showing the distribution of stereom fabrics. Arrows indicate suggested lines of descent (no scale).

to the plate, providing a strong protective crust. Tubercles grow on top of this layer.

Plates with the structure outlined above only evolved in the Mesozoic. Most Palaeozoic echinoids had tests made up of imbricate plates, a condition which is still found in living echinothurioids. Plates are embedded in a thick layer of connective tissue and overlap like the tiles of a roof. There are no deeply penetrating sutural collagen fibres and so no galleried stereom layer is formed. The earliest echinoids had plates that were no more than a thin sheet of undifferentiated laminar stereom (Fig. 3.4). By the Devonian, interambulacral plates had developed an outer layer of dense perforate stereom, but otherwise remained simple. Amongst Palaeozoic echinoids, only palaechinoids evolved tesselate plating and their plates are thick and composed of rectilinear stereom.

All post-Palaeozoic echinoids have coronal plates with at least two layers. In cidarids the outer of the two principal layers is composed of rectilinear stereom whereas in acroechinoids this layer is composed of galleried stereom. Although echinothurioids have imbricate plating, individual plates are composed of two layers, the upper being a laminar stereom (Fig. 3.4). This suggests that echinothurioids may have evolved from a group whose coronal plates were rather more firmly sutured together and that the strong imbrication that we see today is a secondary development.

3.1.4 Resisting impact loading

The evolution from imbricate to sutured plating was an important advance as it increased the shock-resistance capabilities of the skeleton. In a rigid test, impact loading can be transmitted rapidly to adjacent plates thus dissipating the stress. Imbricate skeletons cannot transmit impact shocks in this way and the evolution of sutured plating was a significant innovation allowing post-Palaeozoic echinoids to invade highly turbulent shallow-water habitats.

For energy to be transmitted efficiently from one plate to the next, the stereom of adjacent plates must be in firm contact. Sutural collagen fibres bind the two plates together and, in the simplest case, short stereom knobs extend from each plate to interlock across sutures. The combination of collagen fibres and interlocking calcite is so efficient at transmitting energy across sutures that impact fractures in living animals always occur through the middle of plates and never along sutures. Contact between adjacent plates has been improved in the temnopleurids where peg and socket structures are developed along suture faces (Fig. 2.5).

Resistance to impact loading is most highly developed in the clypeasteroids. Sand-dollars are often found living in the breaker zone of beaches where the pounding by waves can pose a real threat. In primitive groups the plates are connected by small interlocking pegs, but in more advanced groups there are deeply penetrating wedges or thorns to ensure that impact energy is transmitted rapidly with maximum efficiency. Furthermore, clypeasteroids have evolved internal supports (Fig. 3.5). These are firmly sutured together so that impact loading can be transmitted not only to adjacent plates on the same surface but also through the pillars to the opposite side. The pillars are analogous to I-beams used by engineers. Such strengthening is crucial to sand-dollars since without pillars their low,

Figure 3.5 Cross section through part of the Recent sand-dollar *Encope emarginata*, dorsal surface uppermost. The sandwich structure of the test with outer dense layers and a middle spongy layer of irregular pillars provides a lightweight arrangement that is very resistant to bending stress (× 2).

almost flat test would be very much weaker than a globular test of similar diameter.

Finally the spines themselves probably help to reduce the risks of impact loading (Strathmann 1981) since impact energy can be absorbed in the fracturing of spines lessening the likelihood of structural damage to the test.

3.1.5   Ambulacral plate compounding (Fig. 3.6)

Cidarids and all Palaeozoic echinoids have **simple** ambulacral plating – each plate has a single ambulacral pore and one or more small tubercles that do not encroach on to adjacent plates. Most euechinoids, however, have **compound** ambulacral plates where two or more ambulacral plates become fused together during growth and are straddled by a single large tubercle. A third type of ambulacral plating, termed **pseudocompound**, is developed in echinothurioids and some irregular echinoids. Here

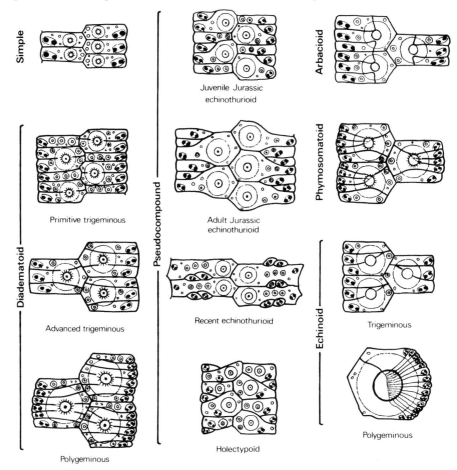

Simple

Juvenile Jurassic
echinothurioid

Arbacioid

Diadematoid

Primitive trigeminous

Pseudocompound

Adult Jurassic
echinothurioid

Phymosomatoid

Advanced trigeminous

Recent echinothurioid

Echinoid

Trigeminous

Polygeminous

Holectypoid

Polygeminous

every third plate is expanded to accommodate a large tubercle and spine at the expense of the other two plates. In Jurassic echinothurioids the two smaller plates of the triad are relatively large and excluded only from reaching the perradial suture. In Cretaceous to Recent echinothurioids, pseudocompounding is much more extreme and the two smaller plates are tiny.

The simplest form of compound plating, diadematoid compounding, first appeared in the late Triassic. In this arrangement all the plates are similar in size and all reach the perradial suture. In all of the early groups, just two of the three plates in a triad were straddled by a large tubercle but in advanced diadematoids all three plates are bound together and overgrown by a large tubercle.

From simple diadematoid compounding several different types of ambulacral plating evolved. Primitive irregular echinoids (pygasteroids and holectypoids) retained a form of diadematoid compounding, but many later groups reverted to simple ambulacral plating since they no longer required large ambulacral tubercles and spines. Pseudocompounding evolved independently several times in irregular echinoids where increased numbers of tube feet were required (e.g. in the

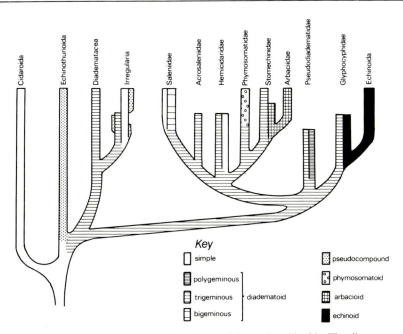

Figure 3.6 Types of ambulacral plating in post-Palaeozoic echinoids. The diagrams on the left-hand side show the principal types of ambulacral plating. The tree on the right-hand side shows the distribution of ambulacral plating amongst post-Palaeozoic echinoids and its possible evolutionary development.

phyllodes of cassiduloids, the petals of clypeasteroids and the frontal ambulacrum of funnel-building spatangoids).

The most complex styles of plate compounding evolved in the echinaceans. The early echinaceans such as pseudodiadematids and acrosaleniids all have plate compounding in the primitive diadematoid manner. In many later groups, one or more of the plates in each set became occluded from the perradial suture through retarded growth. Stomechinids and their descendents, the arbaciids, have compound plates that are still composed of three plates (**trigeminous**), although upper and lower plates in each triad are much smaller and usually occluded from the perradial suture. **Polygeminous** compounding, where more than three plates are united into a compound plate, has evolved independently several times and is found in diadematoid, phymosomatid and echinoid styles of compounding (Fig. 3.6). Finally there are some species, particularly some saleniids, that have returned to having simple ambulacral plates only.

As Kier (1974) has pointed out, the functional significance of plate compounding was that it enabled larger ambulacral tubercles and spines to develop without necessitating any reduction in the number of tube feet. Larger ambulacral spines were required to give the echinoid a more uniform array of spines for both locomotion and defence. Finally, a straight line of ambulacral pores forms a major line of weakness on the test, much akin to the perforations on a sheet of postage stamps. The test can be greatly strengthened by offsetting the ambulacral pores in each compound plate, a strategy that has evolved in several lineages.

3.1.6   Spines and tubercles

Spines are very important appendages used both for locomotion and defence. Each spine is attached to its tubercle by a double ring of tissue, an inner ring of collagenous 'catch apparatus' and an outer ring of muscle (Fig. 3.7). The muscle is quick acting and used to move the spine, whereas the catch apparatus is slow acting and has the ability to contract and hold the spine rigidly in one position. The catch apparatus inserts into the **boss**, which is usually composed of galleried stereom, and muscle attaches to a surrounding platform of retiform or fine labyrinthic stereom termed the **areole**. In perforate tubercles there is an additional central ligament fixing the spine to its tubercle.

Where the areole is uniformly developed around the boss (Fig. 3.8a, b) then the spine muscle can act with equal strength in all directions and the spine has no preferred direction of stroke. The spines of irregular echinoids often have to work in one particular direction against the sediment. To increase the power of the spine stroke in that direction, the areole is enlarged on that side of the tubercle (to accommodate more muscle fibres) and is displaced outwards (to increase the mechanical advantage of the system) (Fig. 3.8c, d). From this it is possible to determine the direction of

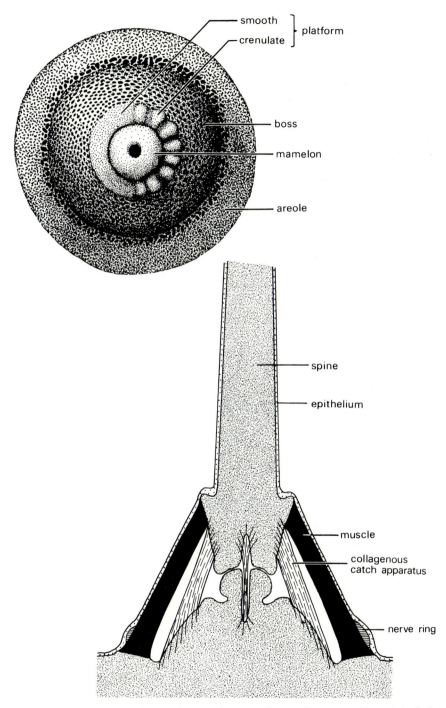

Figure 3.7 Tubercle and spine attachment. The upper diagram shows general morphological features of a tubercle; the lower diagram is a cross section through a tubercle and part of its spine with connecting soft tissue.

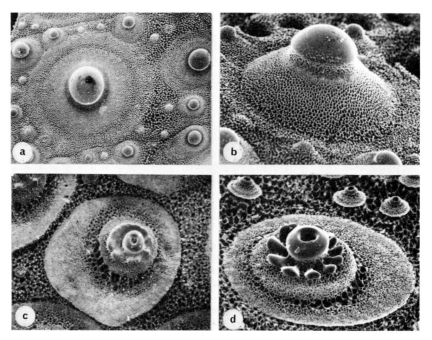

Figure 3.8 Scanning electron micrographs of tubercles. (a, b) Top and side views of symmetrical tubercles (both from *Echinostrephus molaris* (Recent) × 40 and × 80). (c, d) Top and side views of tubercles with a pronounced areole enlargement in one direction – the direction of the spine's action stroke. In both cases the spine does not attach perpendicularly but leans in the direction that the platform is stoutly crenulated or ridged. (c) latero-oral interambulacral tubercle of *Echinocardium cordatum* (Recent) × 40; (d) aboral tubercle from ambulacrum III of *Spatangus raschi* (Recent) × 90.

the power stroke and hence the function of spines in fossil echinoids simply by examining the tubercles (Smith 1980a).

Furthermore, the platform that surrounds the mamelon may be smooth, crenulate or ridged (Fig. 3.7). Crenulation around the platform is matched by a corresponding crenulation around the base of the spine. These interdigitate like cogs when the spine is tilted and help to hold the spine firmly in position. In irregular echinoids, spines that move with a 'rowing' action need crenulation only during that part of the sweep when the spine is actually pushing against the sediment. Crenulation of the associated tubercle is therefore distinctly asymmetric (Fig. 3.8c, d) and this can be used to interpret how spines were held and moved in fossil echinoids.

Some echinoids have large tubercles that are sunken into the plate so that only the mamelon appears above the level of the plate. Sunken tubercles take up less space than tubercles of a similar size built above the plate surface and so are an adaptation for increasing the overall density of spines.

To function efficiently, spines must be light yet strong. They need to be light so that little energy need be expended in moving them and must be strong to resist the bending stresses they are subjected to. Hollow spines provide an excellent compromise since they have the ideal cross section for resisting bending stress, with all their skeletal material distributed in a peripheral ring. Hollow spines are found in many Palaeozoic echinoids and in echinothurioids, diadematoids and irregular echinoids. Indeed, the spines of irregular echinoids are probably more active and subjected to more bending stress than spines of any other group. Cidarids and echinaceans have, however, opted for stronger and more massive spines and have independently evolved solid spines. Diadematoids, despite their formidable array of hollow spines, are preyed upon by large numbers of fish and it may be that solid spines evolved as a deterrent to such predators at the expense of mobility.

Spines are frequently subjected to abrasion and so their external covering of epithelium has to be protected. Most spines are built up of a ring of calcite wedges (Fig. 3.9) so that the nuclei of epidermal cells can lie protected between the wedges. In some groups the external epithelium has been lost altogether and the full-grown shaft is covered by a strong polycrystalline crust.

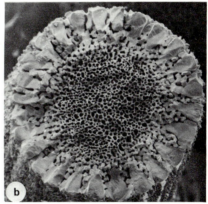

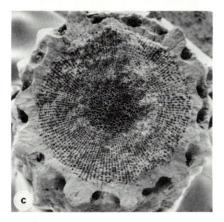

Figure 3.9 Scanning electron micrographs of spines in cross section: (a) hollow spine of the diadematoid *Diadema antillarum* (Recent) × 40; (b) spine of the echinacean *Arbacia lixula* (Recent) – an irregular meshwork forms the core and is surrounded by dense calcite wedges, × 60; (c) spine of the cidarid *Eucidaris tribuloides* (Recent) with a dense polycrystalline cortex, × 25.

3.1.7   Tooth structure

Nowhere is the correlation between function and mechanical design seen better than in the teeth. Teeth are structurally complex and are built up of a large number of different elements (Fig. 3.10). First to form at the growing tip (plumula) of the tooth are the paired primary tooth plates. These are initially triangular but become more elongate as growth proceeds. Side plates then start to form as adaxial expansions to the primary plates. Prisms and laths of stereom also start to appear between primary and side plates to form the prism zone. As growth continues, the tooth plates become welded together by pillars and the prisms become encased in polycrystalline calcite to form a dense, multi-fibred material of high strength. Lying between the primary tooth plates and the prism zone is the stone zone, a zone of very fine needles set in a polycrystalline matrix that is exceedingly strong.

Primary and side plates are inclined towards the centre and form a series of nested cones. In function this is analogous to the self-whetting structure of chisels and it evolved to ensure that the tooth remained ever sharp. The stone zone is a very hard, narrow zone that forms the cutting edge of the tooth. This is supported on either side by the primary and side plates. As the stone zone is worn down, stress is applied to the lowest pairs of plates which eventually shear off in their entirety leaving a new sharp cutting edge (Fig. 3.10). Tooth plates periodically shear off as the cutting edge wears down so that the tooth remains constantly sharp.

The teeth of regular echinoids are used to pluck or rasp at hard surfaces and so are subjected to strong bending stress. As teeth are always drawn in the same direction (towards the centre) the abaxial zone is subjected to compressive stress whereas the adaxial zone is subjected to tensile stress. The structure of the tooth is perfectly adapted to withstand this (Märkel *et al.* 1971). The teeth have a mechanically strong cross-sectional shape analogous with either U- or T-girders. Furthermore, the abaxial part of the tooth is composed of tooth plates separated by pillars, an arrangement best suited for resisting compression, whereas the adaxial part is composed largely of longitudinal prisms set in a matrix, an arrangement ideally suited for resisting tension. The teeth of modern regular echinoids are highly sophisticated structures well adapted for resisting the high bending stresses they are subjected to.

3.2   Adhesion, locomotion and burial

All echinoids are vagile and, with the exception of one or two rock-boring species that never leave their burrows, must constantly move in search of food. Regular echinoids live epifaunally and wander over the sea floor, sometimes climbing rock faces, reef framework or free-standing plants or

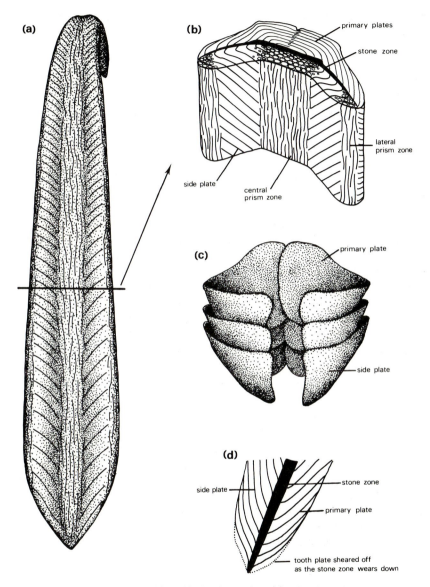

Figure 3.10 Tooth structure and its self-whetting design: (a) axial view of a grooved tooth; (b) block diagram showing cross section as indicated; (c) slightly exploded view of three pairs of tooth plates; (d) cross section through the chewing tip of the tooth showing how it remains ever sharp as it is worn down. (a–c based on the Lower Jurassic pedinoid *Diademopsis*.)

algae. Most irregular echinoids, however, burrow within sediment to avoid predation and gain stability in currents. They are usually highly specialised for their infaunal mode of life.

Euechinoids have specialised balance organs termed **sphaeridia**. These are tiny, swollen spines situated in pits on the more adoral ambulacral plates. (In some sand-dollars the sphaeridia lie completely enclosed within the plates.) The echinoid can ascertain its orientation from the way in which the sphaeridia hang under gravity.

The structure of spines, tubercles and ambulacral pores can provide information on aspects of locomotion and burial in fossil echinoids and it is these features that will be dealt with here.

3.2.1 Adhesion in regular echinoids

Regular echinoids can move in all directions with equal ease and show no preferred orientation. They walk over sediment using only their oral spines, but on hard substrata their tube feet are also brought into play. All oral tube feet of regular echinoids end in a suckered disc and this allows them to climb up rocky faces or free-standing plants and algae.

A typical suckered tube foot is a cylindrical tube with a central fluid-filled lumen (Fig. 3.11). The walls of the stem are made up of an outer sheath of longitudinal and circular connective tissue and an inner bundle of longitudinal muscle fibres. The tube foot extends by contracting its internal ampulla and pumping fluid into the lumen of the tube foot, and contracts by contracting the longitudinal muscle fibres and forcing the fluid back into the ampulla. The suckered disc at the distal end of the tube foot is supported by a skeletal rosette, usually composed of five identical pieces bound together with additional spicules. Muscles run from the edge of the rosette to the centre of the disc. By contracting these muscles the centre of the disc can be pulled inwards making the disc concave and creating a vacuum. Contraction of the stem retractor muscles reinforces the vacuum by forcing the outer edges of the disc upwards to ensure a good seal. The vacuum is broken by relaxing the muscles and.pumping fluid into the tube foot.

The power of adhesion exerted by an echinoid depends not only upon the number of suckered tube feet that it has but also upon the strength of the individual tube feet. The strength of a tube foot depends upon two factors – (i) the strength of the vacuum created, which is proportional to the size of the disc, and (ii) the tensile strength of the stem. In experimental systems the stem almost always ruptures before the suckered disc fails and so the amount of muscle in the stem determines the adhesive strength of the tube foot.

Tube feet are unknown in fossil echinoids, although tube foot rosettes have occasionally been reported (Hess 1973). The structure of the ambulacral pore can, however, give some clue to what the tube feet were

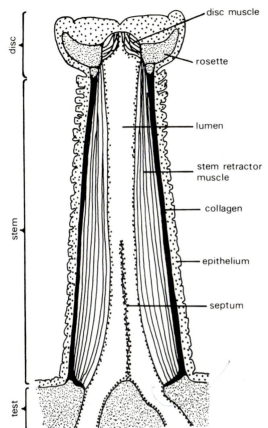

Figure 3.11 Diagrammatic cross section through a suckered tube foot of a regular echinoid.

like. A typical ambulacral pore of a regular echinoid has two holes that connect the tube foot to its internal ampulla (Fig. 3.12). The more per-radial hole is grooved to house a branch of the radial nerve. On the outer surface of the plate the two holes are rimmed by an area of fine-mesh stereom. This is the attachment area for the stem retractor muscle of the tube foot. The width of this area gives a measure of the thickness of stem retractor muscle present and hence an idea of the strength of the tube foot. Ambulacral pores with negligible attachment areas (i.e. less than 20 μm in breadth) bear suckerless tube feet that are sensory or respiratory in function. The strongest tube feet have prominent attachment areas over 200 μm broad.

Oral tube feet are suckered in all living regular echinoids and probably in all but the earliest fossil regular echinoids. Suckered tube feet allow echinoids to climb as well as to clamp the lantern firmly against the sub-stratum for more efficient rasping. It is probably no coincidence that echinoids with many relatively strong tube feet (with attachment areas of 100 μm or more) all have keeled teeth and feed by rasping. Strong tube

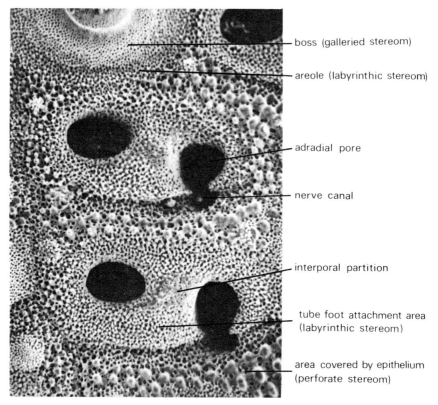

boss (galleried stereom)

areole (labyrinthic stereom)

adradial pore

nerve canal

interporal partition

tube foot attachment area
(labyrinthic stereom)

area covered by epithelium
(perforate stereom)

Figure 3.12 Scanning electron micrograph of two ambulacral pores of *Sphaerechinus granularis* (Recent). The variation in stereom reflects differences in the type of tissue attaching ($\times$ 80).

Figure 3.13 Oral view of part of the test of *Arbacia lixula* ($\times$ 2). Ambulacra expand adorally to form broad phyllodes containing many large ambulacral pores.

feet are also crucial for echinoids living in turbulent shallow-water habitats where they must maintain a powerful grip against wave action. Regular echinoids have adapted for life on exposed rocky coasts by developing phyllodes (Fig. 3.13). These animals have greatly increased the number of their tube feet on the oral surface and, as expected, all ambulacral pores in the phyllodes have broad attachment areas.

Several of the regular echinoid species bore into rock (e.g. *Echinometra*, *Echinostrephus*, *Paracentrotus*). The holes are bored by the action of the spines and the constant rasping of the lantern. They are often parallel sided but may taper slightly towards the aperture. As trace fossils they should be relatively common in ancient beach rocks and have been reported from Middle Jurassic hardgrounds by Palmer (1982).

3.2.2 Locomotion in the irregular echinoids

Recent irregular echinoids live almost exclusively in areas of unconsolidated sediment and use only their spines in locomotion. They are adapted for this mode of life in several ways. First the oral surface has become flatter to bring more spines into contact with the bottom and increase the efficiency of locomotion. The low test profile of most irregular echinoids also helps make them more stable in currents since, unlike the regular echinoids on firm substrata, they cannot adhere to the bottom using tube feet.

Secondly, except for eognathostomates, irregular echinoids move in one direction only. They can only progress with ambulacrum III leading and will rotate if they encounter an obstacle. With unidirectional locomotion the oral spines have developed a power stroke towards the posterior. Fossil irregular echinoids that moved unidirectionally can be easily recognised from their tubercle structure, since the muscle platform (areole) of oral tubercles is enlarged posteriorly to increase the power of the spine stroke in that direction (see p. 34). Unidirectional locomotion enabled the locomotive spines to develop a more efficient beat and led to irregular echinoids employing more systematic detritus-feeding strategies.

Thirdly, as spines are used in many different activities, there is usually some division of labour with functionally and structurally distinct areas of spines and tubercles. This is most pronounced in spatangoids and clypeasteroids (Fig. 3.14), where only a small proportion of the oral spines may be used in locomotion. In spatangoids and holasteroids it is the plastron spines that provide the thrust, whereas in many sand-dollars only the interambulacral spines are used in locomotion. Furthermore, the spines of spatangoids and holasteroids that work against the sediment end in expanded and flattened tips to improve their leverage. Spatulate-tipped spines are best developed in echinoids living in fine sands and muds and least well developed in echinoids living in coarse shell gravels.

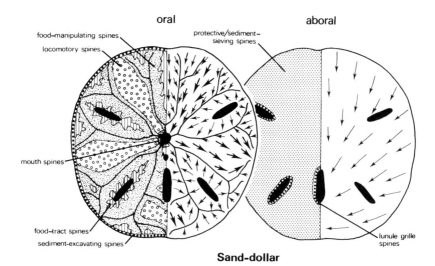

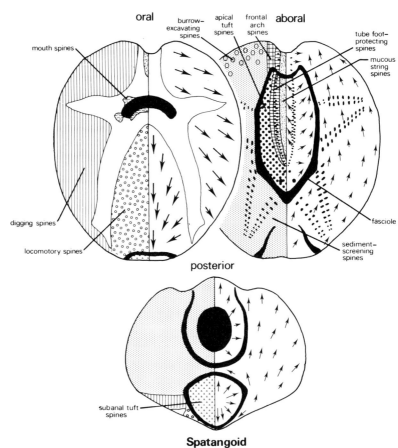

Figure 3.14 Tubercle and spine differentiation in irregular echinoids. Ornamentation on the left-hand side of each diagram indicates the distribution of functionally and morphologically different spines and tubercles. Arrows on the right-hand side indicate the direction of areole enlargement. The sand-dollar is based on *Mellita quinquiesperforata* and the spatangoid on *Echinocardium cordatum*.

3.2.3   Burrowing in the irregular echinoids

Most irregular echinoids live infaunally within unconsolidated sediment. By adopting this mode of life many potential predators can be avoided, and there is much less chance of being overturned or washed out in turbulent conditions when living beneath the zone of mobile sediment. However, living infaunally poses several problems that have to be overcome. Infaunal echinoids require a continuous circulation of oxygenated water past their tube feet. This is no problem in coarse, permeable sediments but becomes difficult in fine grained sediments and so the echinoid needs to maintain a water-filled space around itself to avoid suffocation. Infaunal echinoids also need some mechanism for burrowing into, and cutting through, cohesive sediment. Different groups have overcome these problems in different ways.

There are three main burrowing techniques employed by echinoids. In eognathostomates the oral tubercle arrangement suggests that all oral spines had a radially directed power stroke (Fig. 3.15a). They presumably dug vertically into the sediment by pushing material radially out from underneath themselves. In contrast, most cassiduloids and clypeasteroids burrow by ploughing forwards into the sediment. The sediment that piles

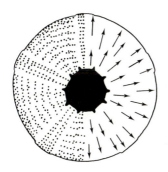

(a) Sediment excavated radially

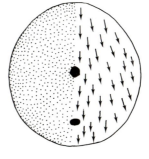

(b) Ploughs into sediment

(c) Sediment excavated laterally

Figure 3.15 Burrowing strategy and oral tubercle arrangement in irregular echinoids. Tubercle density and arrangement is schematically shown on the left-hand side while arrows on the right-hand side show the direction of areole enlargement.

up in front of the test is lifted aborally by the tube feet and spines and this is then transported posteriorly by the aboral spines. Thus oral spines are not used to excavate sediment but simply provide the forward thrust, and so their tubercles all have a posterior areole enlargement (Fig. 3.15b). The third method of burrowing is developed in atelostomates as well as in cassiduloids such as *Cassidulus* and probably in fossil clypeasteroids such as *Lenita*. In these echinoids there are two lateral zones of oral spines that lie almost flattened against the test and excavate sediment from beneath the animal laterally with a rowing action. They sink into the sediment with little forward motion. Oral tubercles in these zones have a pronounced areole enlargement in a lateroposterior direction (Fig. 3.15c). Furthermore, because the spine is far from perpendicular to the test, the boss and mamelon of tubercles are typically asymmetric.

Burrowing into the sediment is just the first of the problems. Many irregular echinoids move continuously forward through the sediment. In loose, unconsolidated sediments no particular adaptations are needed but, in more cohesive sands and muds, burrowing is made easier by having a test that is wedge-shaped in profile, and by having at least some of the anterior facing spines specialised for excavating sediment from the front wall of the burrow. Sediment-excavating spines are generally short, stout and pointed and their tubercles are larger and readily distinguished on the test (Fig. 3.16).

An infaunal echinoid must also maintain a flow of oxygenated water

Figure 3.16 Tubercles that support spines of different function are usually very different in structure. The large central tubercle with the crenulated platform supports a spine used for excavating sediment from the front wall of the burrow. The small surrounding tubercles bear curved and spatulate-tipped spines that hold the sediment away from the surface of the test. (Anterior interambulacral tubercles of the spatangoid *Echinocardium cordatum* (Recent), × 60, from Smith 1980a, courtesy of the Royal Society.

through its burrow and keep sediment away from the surface of the test. In the simplest cases an envelope of water is maintained around the animal by its spine canopy. Aboral spines of infaunal echinoids are uniform in size and evenly spaced so as to form a grille of sufficient density to prevent sediment particles from falling between the spines. In this way a fluid-filled space is maintained between the tips of the spines and the test surface. As might be expected, there appears to be a correlation between aboral tubercle density (spine canopy density) and the grade of sediment inhabited (Fig. 3.17). Irregular echinoids with a sparse scattering of aboral tubercles, of various sizes, are almost certainly epifaunal and aboral tubercle density increases in finer-grained sediments. However, there is a practical limit to the density of spines that can be achieved (the highest density being about 15 tubercles per mm^2) and so the spine canopy alone cannot provide a barrier to silt- or mud-grade sediments. This explains why cassiduloids, clypeasteroids and holectypoids are all restricted to burrowing in sands or gravels. In fine-grained sediments other adaptations are required to maintain the water envelope around an infaunal echinoid.

Today, only spatangoids can successfully cope with living in silt or mud, although in the past certain holasteroids may also have managed to live in such sediments. They maintain a water-filled space around themselves by producing a sheath of mucus held by the spine tips and coating the entire dorsal surface. This forms a screen that prevents even very fine sediment particles from falling between the spine canopy. The spine canopy itself is more efficient in spatangoids than in other irregular echinoids because the aboral spines are curved and overlap, and they all end in an expanded spatulate tip. The mucous coat is secreted by the highly specialised spines of the dorsal fasciole in all living spatangoids, but may possibly have been

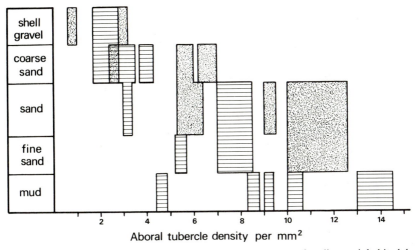

Figure 3.17 Aboral tubercle density plotted against the type of sediment inhabited by infaunal irregular echinoids. Data from 24 species (cassiduloids, clypeasteroids and holectypoids stippled; spatangoids lined) redrawn from Smith (1980a).

secreted by all dorsal miliary spines in primitive groups such as the micras-terids. A comparable situation can be found in the living sand-dollar *Echinarachnius*, which has a very dense canopy of mucus-secreting miliary spines dorsally and has been reported from silty sediments.

Fine-grained sediments are generally too impermeable to allow suffi-cient water to circulate through the burrow. In fine sands and muds it becomes crucial for infaunal spatangoids to construct a shaft up to the surface so that water can be drawn into the burrow. This shaft is excavated by specialised tube feet situated in the more adapical part of ambulacrum III. The ambulacrum in this region is usually broad and sunken, and there is a tuft of large spines from the immediately adjacent interambulacra to maintain the lower part of the shaft and give the funnel-building tube feet room to manoeuvre. Water is actively drawn into the burrow by ciliary currents which are produced by the spines of the fascioles. In coarse, perme-able sediments spatangoids do not need to construct a shaft to the surface as water can be drawn through the sediment interstices by ciliary currents.

Infaunal echinoids also face the problem of having to get rid of the water that they draw into their burrow, so as to maintain a continuous inflow of oxygenated water. Many spatangoids construct a posterior tunnel or pair of tunnels to the burrow and Chesher (1968) suggested that these tunnels act as drainage tubes, increasing the surface area of the back wall and so helping to rid the burrow of its excess water. These subanal tunnels are constructed by specialised subanal tube feet. There is usually also a single or double tuft of spines in this region and a fasciole to help pump water backwards into the back wall of the burrow.

When trying to decide whether or not a particular fossil spatangoid or holasteroid burrowed, the following morphological features are useful guides:

(a) A dense, uniform development of aboral tubercles is indicative of an infaunal mode of life. Those with a variably sized array of sparse tubercles are typically epifaunal or are restricted to living in coarse-grained sediments.
(b) The presence of distinctly larger oval ambulacral pores in the adapical region of ambulacrum III compared to ambital ambulacral pores usu-ally indicates that funnel-building tube feet were present.
(c) A sunken anterior ambulacrum with enlarged interambulacral tuber-cles immediately adjacent is characteristic of heart-urchins that burrow in sands or muds and construct shafts to the surface. Where ambulac-rum III is narrow and lies flush with the rest of the test and has neither enlarged interambulacral tubercles nor larger adapical pores, the heart-urchin lives either epifaunally or within coarse sand or shell gravel without constructing a vertical shaft.
(d) Aboral fascioles (i.e. peripetalous, apical or lateral fascioles) may or may not be present in heart-urchins that burrow in coarse sands or

gravels but are always present in those that burrow in fine sands and muds.

(e) The presence of subanal tunnel-building tube feet is easily recognised in fossils because their ambulacral pores are always much larger than adjacent sensory tube-foot pores. They may or may not be enclosed by a subanal or latero-anal fasciole. Some spatangoids can burrow even in muds, without constructing a subanal tunnel, while some semi-infaunal species have both tunnel-building tube feet and a fasciole.

3.2.4   Trace fossils

Although until quite recently fossil burrows of irregular echinoids were thought to be rare trace fossils, it is now evident that bioturbation traces of heart-urchins are relatively common from the Cretaceous onwards. The most detailed studies of echinoid bioturbation are those of Bromley and Asgaard (1975), describing Pliocene traces produced by the spatangoid *Echinocardium*, and Smith and Crimes (1983) who argue that the common trace fossils *Scolicia* and *Subphyllochorda* were produced by heart-urchins.

Burrows of heart-urchins consist of a cylindrical core of sediment with curved, backfill laminae, a much smaller single or double tube of sediment (the subanal tunnels) either within or at the base of the backfilled core of sediment, and a tripartite lower surface consisting of outer zones bearing fine oblique scratch traces and a central zone that may have prod traces (Fig. 3.18). Many of these belong to the ichnogenus *Subphyllochorda*.

Top surface traces in negative relief belonging to the ichnogenus *Scolicia* have the same tripartite structure and backfill laminae as *Subphyllochorda*, and have been interpreted as surface furrows formed by heart-urchins ploughing semi-infaunally along the sea floor.

Resting traces, where an individual has burrowed deeper for temporary refuge, belong to the ichnogenus *Cardioichnus*. These are ovoid sole marks preserved in positive relief at sandstone–shale boundaries. They show lateral scratch traces of sediment-excavating spines and a central V-shaped region with prod traces that appears to correspond with the plastron (Fig. 3.18).

3.2.5   Evolutionary history

Throughout the Palaeozoic all echinoids appear to have lived epifaunally, probably favouring firm or stable bottoms in quiet offshore or protected habitats. It was not until the Mesozoic radiation that echinoids diversified and adapted for different habitats. Cidarids and diademataceans continued to occupy a similar niche to their Palaeozoic ancestors, living in shallow, protected habitats such as lagoons, or ranging into deeper offshore waters. The early echinaceans started off in similar habitats but quickly diversified, so that several groups adapted for life in shallow turbulent habitats on reefs

Figure 3.18 Trace fossils produced by heart urchins. (a) *Subphyllochorda*, a trail in positive relief on a sandstone sole surface probably produced by an infaunal heart urchin (× 1). (b) *Cardioichnus*, three resting traces (one continuous with *Subphyllochorda*) in positive relief on a sandstone sole surface produced where heart urchins have temporarily burrowed more deeply (× 1). (From Smith & Crimes 1983, courtesy of the International Palaeontological Society.)

or rock platforms. They achieved this by developing more numerous and stronger oral tube feet, and the first echinaceans with well developed phyllodes appeared in the Middle Jurassic.

The first irregular echinoids appeared in the Lower Jurassic but these pygasteroids were ill adapted for burrowing and probably lived epifaunally. Evolution proceeded rapidly and, before the end of the Lower Jurassic, both holectypoids and galeropygoids had developed a denser array of uniformly sized dorsal spines and were able to live infaunally and to invade areas of coarse mobile sediment. Galeropygoids also show a recognisable areole enlargement to the posterior of oral tubercles, and appear to have been the first echinoids to have unidirectional locomotion and a more organised feeding strategy. Disasteroids and cassiduloids both evolved from galeropygoids and had a similar dorsal spine density, which suggested that they were probably unable to live within fine sands or muds. However, each gave rise to a group which could live in finer sediments. Disasteroids gave rise to holasteroids and spatangoids in the Cretaceous, and groups evolved with funnel-building tube feet, dorsal fascioles and a dense screen of spatulate aboral spines. It seems that several groups of spatangoid and possibly some holasteroids independently adapted to live infaunally within fine-grained sediment once the crucial morphological innovations had appeared. Finally, in the early Tertiary, clypeasteroids evolved from cassiduloid ancestors and, by developing tiny but very dense aboral spines, were able to live within fine sands that had been previously unavailable to cassiduloids.

3.3   Feeding and waste disposal

3.3.1   Feeding strategies

The development of new food-gathering techniques has been one of the most critical factors in the evolution of echinoids. The earliest echinoids were probably detritus feeders but the group has since diversified to utilise a wide variety of foods. Fortunately, most changes in diet have been accompanied by specific morphological adaptations, so that it is often possible to make deductions about feeding strategies of fossil echinoids.

In broad terms, living echinoids can be described as either 'scrapers' or 'sediment swallowers' depending upon how they feed. The regular echinoids are predominantly scrapers and use their dental apparatus to rasp, pluck or scoop up suitable material. Irregular echinoids, on the other hand, are usually microphagous, feeding on the tiny organic particles found on and amongst the grains of sediment. They may have highly specialised techniques for gathering their food. A recent review of biological aspects of feeding in echinoids can be found in Jangoux and Lawrence (1982).

3.3.2   Feeding in regular echinoids

Regular echinoids are remarkably generalised in their diet and will feed on a variety of different things (Lawrence 1975). Availability seems to be the most important factor governing what is eaten. None is restricted to any one particular food though many show some preference in diet. Regular echinoids are able to exploit four types of food: (a) attached or drifting plants and algae; (b) encrusting and boring organisms (e.g. *Cliona*, spirorbids, bryozoans, algae); (c) sessile organisms (e.g. sponges, gorgonids, corals); (d) detritus. Deep-sea species feed largely on detritus and their guts are packed with bottom sediment and whatever drifted plant material there is available. In shallower waters of the continental shelf, regular echinoids feed mainly on sessile or encrusting organisms or on free-standing plants and algae.

Echinoids have some ability to recognise potential food at a distance using chemosensory receptors and will move in the appropriate direction. Once located, the food is 'tasted' by the tube feet, particularly by the large fleshy peristomial tube feet that surround the mouth. The teeth are then brought into play to cut or rasp, and material is passed into the mouth with the help of the peristomial lips.

Manoeuvring plant material to the mouth requires both spines and tube feet working together. This method of feeding is particularly important for echinaceans, such as *Echinostrephus*, which bore into rock. Individuals rarely, if ever, leave their burrows and some may even grow too large to be able to squeeze out (Campbell *et al*. 1973). Although they can graze on the microscopic algae that grow on the walls of the burrow, they have to supplement their diet by capturing larger plant material that floats past. *Echinostrephus* does this by moving to the top of its burrow (still in a horizontal attitude) and extending its aboral tube feet. The suckered tube feet are able to catch and hold suitable material which, with the aid of spines, is then drawn in and passed to the mouth.

Pedicellariae, tiny pincer-like stalked appendages, may also play a part in feeding though they are probably more important as organs of defence (see p. 98). Ophicephalous pedicellariae of regular echinoids and bidentate pedicellariae of clypeasteroids are able to capture and hold live prey which can then be passed to the mouth.

Recently it has been shown that echinoids can absorb dissolved organic material from sea water (Pequignat 1970). Although epithelial absorption of dissolved organic material must provide the major source of nutrients for the external appendages, this method of feeding is relatively unimportant to the animal as a whole.

3.3.3   Feeding in irregular echinoids

In contrast to regular echinoids, the great majority of irregular echinoids are bulk sediment swallowers. The only living group of irregular echinoids to retain a functioning dental apparatus as adults are the clypeasteroids, and their lantern is wholly internal and used for crushing rather than scraping.

The ways in which irregular echinoids have evolved to collect their diet of sediment are quite varied. Amongst living groups, the simplest feeding techniques are found in cassiduloids, holectypoids and primitive clypeasteroids. These animals live in sands or gravels and ingest sediment for the minute organic material that it contains. Oral tube feet are suckered and used to pick up and transfer particles to the mouth. Spines adjacent to the mouth are densely packed and form a grille across the peristome. These help manipulate sediment particles into the mouth.

Spatangoids and many holasteroids have a few highly specialised tube feet around the mouth that are used to collect sediment. These phyllode tube feet end in chimney-brush-like discs (Fig. 3.19) that are prehensile and pick up particles, both large and small, by mucous adhesion. Pourtalesiids, a peculiar group of deep-sea holasteroids with a bottle-shaped test, have abandoned using tube feet to collect sediment. Instead, the mouth lies at the end of a funnel-shaped frontal ambulacrum which is lined with flattened geniculate spines. These spines probably help shovel the top layer of sediment into the mouth as the animal ploughs along the surface.

Many of the deeply burrowing heart-urchins supplement their diet by

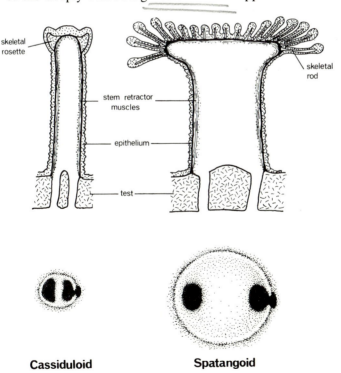

Figure 3.19   Food-gathering tube feet of irregular echinoids and their ambulacral pores. Cassiduloid tube feet work using suction; spatangoid tube feet work using mucous adhesion.

feeding on the flow of fine suspended material that is brought into the burrow and down the frontal ambulacrum by ciliary currents. Indeed, the spatangoids *Echinocardium*, *Moira* and *Schizaster* have developed a mucous string conveyor system in their anterior ambulacrum. Mucus is secreted by the miliary spines in the dorsal part of the frontal groove and is loosely compacted and moved adorally by the action of these spines. Material brought into the burrow by ciliary currents or tube feet becomes incorporated into the string.

Mucus is also used for transporting food in the more specialised sand-dollars that possess a system of food grooves on their oral surface (Fig. 3.20). Like all clypeasteroids, they have an enormous number of minute suckered tube feet which are used to pick up suitable particles. These particles can then be transferred directly into the nearest food groove rather than being transferred adorally from tube foot to tube foot. Sand-dollars feed exclusively on fine detrital material that they sieve from the uppermost layer of sand. They do this in a very elegant way (Ghiold 1979, Seilacher 1979). As they continuously burrow forwards immediately beneath the surface, they pass a thin veneer of sediment over their dorsal surface. The spines are dense enough to prevent sand grains from falling through the spine canopy but finer material passes between the tips of the spines and is transported to the oral surface in a strong ciliary current near the base of the spine shafts. On the oral surface the ciliary currents continue to transport the fine material adorally towards food grooves where it is captured by the tube feet. Tube feet transfer their load to the food grooves where it becomes incorporated into a mucous string that is moved slowly towards the mouth.

There is at least one sand-dollar that has adapted for suspension feeding.

Figure 3.20 Food grooves in clypeasteroids. (a) *Scutella faujasii* (Miocene): food grooves bifurcate close to the mouth and diverge to cover as much of the periphery as possible ($\times$ 0.66). (b) *Mellita quinquiesperforata* (Recent): food grooves have many side branches leading towards the lunules and test margin ($\times$ 0.5).

In moderate currents, the sand-dollar *Dendraster* inclines itself, with only the anterior embedded in the sediment, and feeds on suspended material (Timko 1976). Larger suspended particles are captured by spines. pedicellariae and tube feet, while finer material is drawn into the food grooves by ciliary currents. This method of feeding is used only in suitable flow regimes. In more tranquil areas *Dendraster* lives horizontally and feeds in the more usual manner.

3.3.4 *Morphological adaptations*

Morphological changes, associated with the evolution of feeding and excretion, that can be recognised and interpreted in fossil echinoids fall into three broad categories – those affecting overall shape, those affecting the dental apparatus and those affecting tube feet and spines.

Shape   The change of diet from 'scraping' to 'sediment swallowing' that occurred during echinoid evolution affected the shape of the test. Regular echinoids generally have a relatively high organic intake, even those feeding on encrusting organisms. They also have a high absorption efficiency ranging from 30–40% for grazers up to 90% for carnivores (Lawrence 1975). The indigestible portion of the diet is periodically voided through the anus as mucus-bound faecal pellets. Diadematoids rotate their large anal cone in order to force out and disperse the pellets, but dispersing faecal material is rarely a problem for regular echinoids as they live epifaunally and are bathed in currents.

Bulk sediment swallowing poses a much greater problem. The organic content of sediment is often extremely low so that large quantities have to be ingested. Most of this is totally indigestible and is voided, so that irregular echinoids such as the cassiduloid *Apatopygus* have an almost continuous discharge of sediment from the anus (Higgins 1974). As many of these echinoids live infaunally, the discharge cannot simply be washed away by currents. This high rate of faecal discharge prompted evolutionary adaptations to prevent fouling of the aboral respiratory surface and decrease the likelihood of reingestion.

These problems were solved very early on in the history of irregular echinoids with the progressive migration of the periproct towards the posterior and away from the apex (Fig. 3.21). If the group Irregularia is defined on the presence of diamond-shaped teeth, then the most primitive members of Lower Liassic age such as *Eodiadema* and '*Plesiechinus*' *hawkinsi* still have a normal monocyclic apical system enclosing the periproct (Fig. 3.22). From this point Jesionek-Szymanska (1968) has shown that irregular echinoids evolved along two divergent paths, one leading to the holectypoids, the other to all other living irregular groups. In the first lineage, which includes all later pygasteroids, the periproct moved to the posterior of the apical system with the apparent loss of genital 5 and without causing

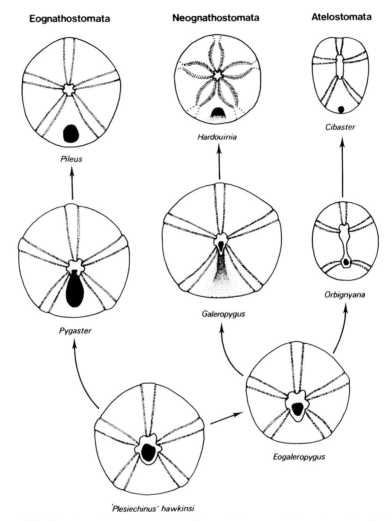

Figure 3.21 Migration of the periproct out of the apical system in irregular echinoids. The apical system is heavily outlined (in *Galeropygus* the periproct opens into a posterior sulcus).

oculars I and V to elongate. In pygasteroids the periproct remained aboral, but in holectypoids it quickly moved to the oral surface, leaving a compact apical system that has incorporated an accessory plate in place of the lost genital plate.

The other lineage passed through a rather different and more gradual transition. Oculars I and V became elongate as the periproct moved towards the posterior and genital 5 was retained, though in a modified form and with no gonopore (Fig. 3.22). In the galeropygoids, the anus remained close to the apex, but lay at the top of a deep posterior groove termed the **anal sulcus** (Fig. 3.21). This groove channelled the faecal

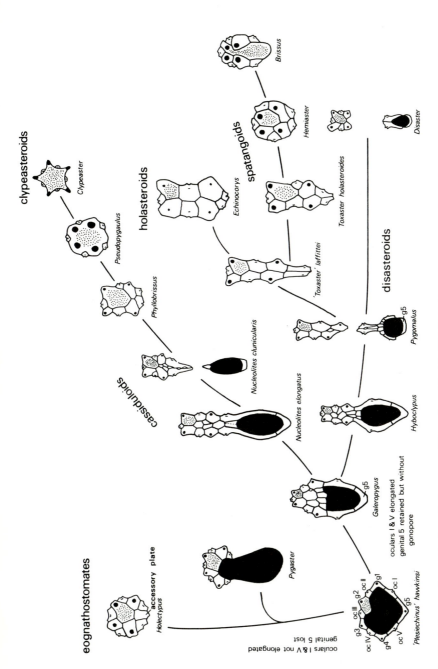

Figure 3.22 Evolutionary changes in the apical systems of irregular echinoids. The lines do not necessarily indicate direct lines of descent but show general trends. Apical systems of individual species taken from Jesionek-Szymanska (1963, 1979), Devries (1960) and Kier (1974): no scale.

discharge backwards away from the aboral respiratory surface. The periproct moved further to the posterior in Jurassic cassiduloids, leaving the elongate oculars I and V still in contact with the apical system. By the Cretaceous, most cassiduloids had their periproct situated well to the posterior and had evolved a more compact tetrabasal apical system, and by the Senonian the great majority had evolved a monobasal apical system with just one large genital plate (Kier 1962). This was inherited by the clypeasteroids where some groups later re-evolved a fifth gonad and gonopore.

An anal sulcus did not develop in disasteroids, but here the migration of the periproct carried genital 5 and the two posterior oculars backwards with it forming a disjunct apical system (Fig. 3.22). Later on, in the Upper Jurassic, oculars I and V tended to migrate back towards the rest of the apical system. This eventually led to the formation of the elongate tetrabasal system of holasteroids and the more compact tetrabasal system of spatangoids.

The critical factor that triggered off these dramatic changes seems to have been the adoption of a fully infaunal mode of life. Early pygasteroids with an apically situated anus probably succeeded as well as they did only by living epifaunally or semi-infaunally. With the change to an infaunal mode of life, faecal discharge became a critical problem and either the anus rapidly become more posterior in position or an anal sulcus was evolved.

The greatly off-centred periproct found in some acrosaleniids represents an independent evolutionary trend within echinaceans. Possibly acrosaleniids had, like irregular echinoids, adopted a sediment-rich diet but had remained epifaunal.

Another change associated with feeding is the development of a more flattened test. This has occurred twice, once in the Upper Palaeozoic and once in the Tertiary. Lower Palaeozoic echinoids mostly have tall and rounded tests but, in the Carboniferous and Permian, proterocidarids evolved much flatter tests and their ambulacra became greatly expanded on the oral surface (Kier 1965). This flattening increased the proportion of tube feet in contact with the bottom, and these tube feet are thought to have been used for food gathering (see p. 76). Clypeasteroids also became very much flatter in the Eocene during the rapid evolution of sand-dollars. Here, however, test flattening served a different purpose, allowing sand-dollars to develop their highly specialised method of detritus feeding. It is only with a flattened test that a thin sheet of sediment can be carried over the dorsal surface and sieved of its fine material. The flattened oral surface that is variably developed in other irregular echinoids is not associated with an obvious increase in tube feet and is an adaptation for more efficient locomotion on unconsolidated sediments.

As might be expected, there have also been some important evolutionary changes in the shape and position of the peristome. Palaeozoic echinoids all have relatively small peristomes, centrally positioned and largely covered by a flexible, plated membrane. The plates of the peristomial membrane imbricate, and there are series of both ambulacral and non-ambulacral plates (Fig. 3.23). Each ambulacral plate is perforate and would have had a sensory tube foot whereas non-ambulacral plates (not to be confused with interambulacral plates) are imperforate and interradial in position. In cidarids, the peristome forms a relatively greater part of the oral surface, correlating with the more mobile lantern, and peristomial plates are reduced to ten columns of ambulacral plates and five columns of non-ambulacral plates. Echinothurioids also have ten columns of ambulacral plates on the peristome but in this group there are no non-ambulacral plates.

In regular acroechinoids the peristome is often relatively large, allowing the lantern considerable freedom of movement. On the peristome, ambulacral plates are reduced to ten large buccal plates each with a sensory tube foot. The rest of the peristomial membrane is covered in tiny non-ambulacral plates (Fig. 3.23). The peristome remains relatively large in primitive irregular echinoids, but with the loss of a functioning lantern a large peristome was no longer needed, and there was a dramatic decrease in the size of the peristome in galeropygoids and their descendents (Fig. 3.24). Even clypeasteroids do not require a large peristome since their lantern is wholly internal. The first-formed ambulacral plates no longer extend on to the peristome but are firmly bound into the corona. The peristomial membrane may include minute non-ambulacral plates, as in spatangoids, or may have no plates whatsoever, as in clypeasteroids.

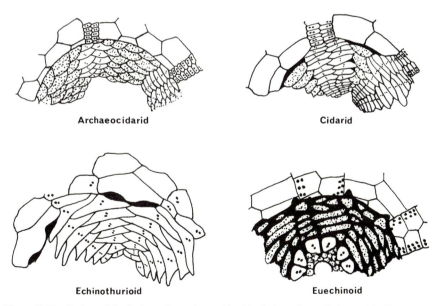

Archaeocidarid   **Cidarid**

Echinothurioid   **Euechinoid**

Figure 3.23 Peristomial plating of regular echinoids (taken from Jackson 1912). Non-ambulacral plates stippled.

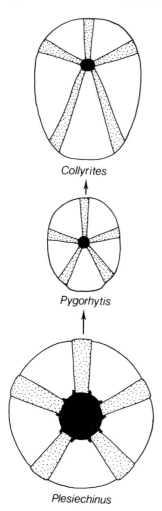

Collyrites

Pygorhytis

Plesiechinus

Figure 3.24 Changes in size and position of the peristome in primitive irregular echinoids. Pygasteroids (*Plesiechinus*) have a large central peristome with buccal slits. Early disasteroids (*Pygorhytis*) have a much smaller peristome without buccal slits while in later disasteroids (*Collyrites*) the peristome lies towards the anterior.

During the evolution of irregular echinoids there is a pronounced tendency for the peristome to shift anteriorly. This is first seen in the galeropygoids, where the peristome may be central (*Eogaleropygus*), slightly anterior to centre (*Galeropygus*) or well anterior (*Hyboclypus*). It is probably no coincidence that this change occurs with the evolution of a strongly marked unidirectional locomotion. In cassiduloids and clypeasteroids the coronal plates surrounding the peristome are turned inwards to form a well to the mouth. This increases the number of spines that arch across the peristome and help to manipulate particles into the mouth.

In disasteroids as well as many holasteroids and early spatangoids, the peristome is circular and flush with the surface of the test. Later, as more reliance came to be placed on food transported down the anterior ambulacrum, the peristome moved even closer to the anterior edge of the

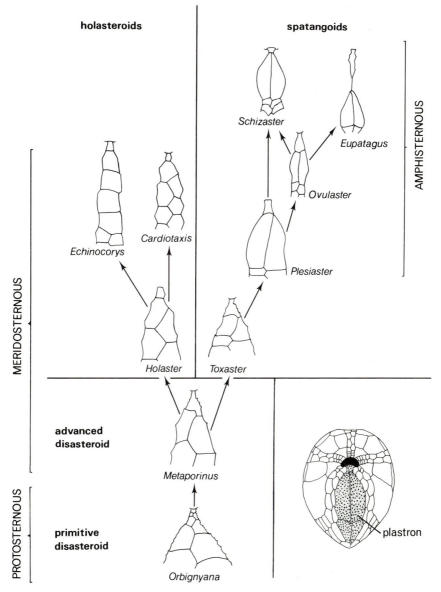

holasteroids

spatangoids

MERIDOSTERNOUS

AMPHISTERNOUS

PROTOSTERNOUS

Schizaster

Eupatagus

Ovulaster

Cardiotaxis

Echinocorys

Plesiaster

Holaster

Toxaster

advanced disasteroid

Metaporinus

primitive disasteroid

Orbignyana

plastron

Figure 3.25 Evolution of plastron plating in the Atelostomata. The adoral plating in the posterior interambulacrum is shown and developed from the protosternous arrangement to the meridosternous and amphisternous arrangement by the Cretaceous. Lines do not indicate ancestor/descendent relationships but show general morphological trends. (Details of plating in disasteroids taken from Jesionek-Szymanska 1963, courtesy of the author.)

ventral surface and became forward facing through the development of an enlarged labrum. Clearly, material collected from directly beneath the mouth was becoming less important.

The changes in shape and position of the peristome could only take place if there were structural changes in the plating of the posterior interambulacrum. These changes have been documented by Jesionek-Szymanska (1963), Mintz (1968) and Kier (1974). Early disasteroids are **protosternous**, that is to say they have no truly developed plastron (Fig. 3.25). Plastrons that have an enlarged labrum followed by two offset plates are known as **meridosternous** – a condition found in several groups of disasteroids and in the more primitive holasteroids and spatangoids. In later spatangoids, the labrum is followed by two opposed plates and the plastron is said to be **amphisternous**.

Finally, the adoption of suspension feeding in the sand-dollar *Dendraster* has led to some important changes in overall design. Because *Dendraster* feeds in an inclined position with the anterior of the test embedded in the sediment, the petals are positioned unusually far back and the distribution of food grooves is correspondingly modified. Anterior food grooves are short and stop well before the margin, whereas posterior food grooves are extensive and may even continue on the aboral surface. As sand-dollars with this distinctive morphology first appear in the Pliocene, suspension feeding appears to be a recent development.

The lantern   The echinoid dental apparatus is a complex internal structure composed of up to 40 elements arranged in five identical units. When fully developed, each unit consists of a pair of hemi-pyramids, a pair of epiphyses, a tooth, a rotula and a two-pieced compass (Fig. 3.26). The lantern is moved by 60 muscles, yet despite this complexity the lantern as a whole functions with elegant simplicity.

Tooth structure has been discussed previously from a mechanical point of view (p. 38). The teeth of living echinoids are highly sophisticated rasping or crushing elements, but these developed in the Mesozoic, and Palaeozoic teeth are much simpler. The teeth lie within and are supported by the **pyramids**, each composed of two large, mirror-image elements that are sutured together interradially (Fig. 3.26). Hemi-pyramids may be sutured together along their entire height, as in cidarids, or may abut for only a small distance, leaving a large aboral notch known as the **foramen magnum**, as in euechinoids. The pyramids usually have a smooth adaxial platform, known as the **dental slide**, on which the tooth rests, and the adoral end of each hemi-pyramid curves round to clasp the tooth firmly. Sutured to the apical end of each hemi-pyramid is an axe-shaped element, the **epiphysis**. Adjacent pyramids are connected by the **rotula**, a flattened brace attached to the epiphyses. The rotulae improve the action of the lantern as a whole, while allowing a limited amount of relative movement between individual pyramids on uneven surfaces. A **compass** lies above

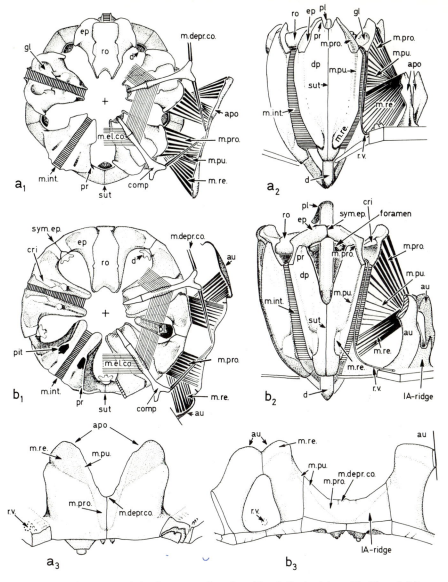

Figure 3.26 Structure of the lantern and perignathic girdle in (a) a cidarid and (b) a camarodont. (1) Apical view; (2) lateral view; (3) perignathic girdle (from Märkel 1981, courtesy of Springer Verlag).

Key

apo	apophysis	m. el. co.	compass elevator muscle
au	auricle	m. int.	interpyramidal muscle
comp	compass	m. pro.	protractor muscle
cri	crista	m. pu.	postural muscle
d	tooth	m. re.	retractor muscle
dp	hemi-pyramid	pl	plumule of tooth
ep	epiphysis	pr	super alveolar process
gl	glenoid cavity	ro	rotula
IA	interambulacrum	r.v.	radial water vessel
m. depr. co.	compass depressor muscle	sut	intrapyramidal suture

each rotula. It is composed of two slender elements sutured together, end to end, and can be raised or lowered to alter the internal pressure of the peripharyngeal coelom, the body space that encloses the lantern, compensating for movement of the lantern in and out of the peristome.

There are a great many muscles required to work the lantern. The muscles that close the jaws are the large and powerful interpyramidal muscles running between the lateral wings of adjacent hemi-pyramids. Contraction of these muscles draws the jaws together. The superior and inferior rotula muscles bind the rotula to its epiphyses and are used to pull adjacent pyramids level after any relative displacement. Protraction, retraction and lateral movement of the whole lantern is accomplished by the muscles that run from the hemi-pyramids to the perignathic girdle. Each pyramid is connected to the perignathic girdle by a pair of protractor muscles, a pair of postural muscles and a pair of retractor muscles, but there is no sharply defined boundary between these muscles. Lantern retractor muscles also open the jaws on contraction. Finally, both circumferential and radial muscles attach to the compasses and are used to raise and lower those elements.

The evolution of the lantern is a fascinating story that has proved most useful in unravelling the phylogeny of echinoids. The earliest lanterns that are known come from two Upper Ordovician echinoids *Aulechinus* and *Ectinechinus*, first described by MacBride and Spencer (1938). Here the lantern is extremely simple, consisting of five broad teeth each supported by a pair of hemi-pyramids and a second pair of more distal plates equivalent to epiphyses. There are no rotulae or compasses as far as can be seen. The lantern is also unusual in being very flat. In *Ectinechinus*, hemi-pyramids are rather narrow, gently curved elements that become distinctly ridged at their adoral end (Fig. 3.27). Each pair of hemi-pyramids appears to be united for only a small part of its length, leaving a large and broad foramen magnum. Laterally, there is a clear muscle attachment platform for the interpyramidal muscles, but this is not developed into a wing-like process as on later pyramids. The adoral ridging may mark the attachment area for jaw-opening muscles. The teeth of *Aulechinus* and *Ectinechinus* are extremely simple, consisting of nothing more than a series of long circular rods bound together to form a broad flat blade (Fig. 3.28). The lantern clearly had strong interpyramidal muscles to close the jaws, but how were the jaws opened? Each hemi-pyramid lies closely associated with the most adoral ambulacral plates and it seems likely that muscles from the pyramids attached to the ambulacral plates and acted as the main jaw openers. As with all Palaeozoic echinoids, there was no perignathic girdle and these muscles must have been small and relatively weak. The lantern functioned as a scoop, not a scraper, and jaw movement must have been largely confined to opening and closure in the horizontal plane.

The lantern of the Lower Silurian *Aptilechinus* is more like present-day lanterns with both rotulae and epiphyses and a rather more upright

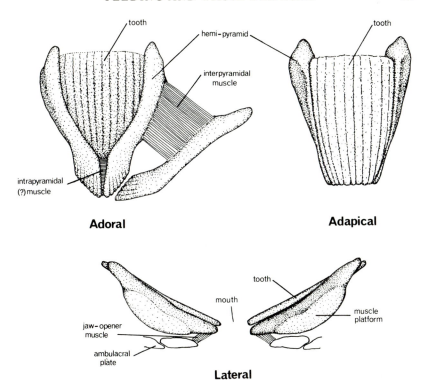

Adoral

Adapical

Lateral

Figure 3.27 Reconstruction of the lantern in one of the earliest known echinoids, *Ectinechinus*, from the Upper Ordovician.

pyramid. Although compasses may have been present, the earliest that have been found come from the Upper Silurian. The appearance of rotulae and compasses in the Silurian suggests that lanterns were becoming more like grabs than scoops, and were more actively protruded and retracted.

The Upper Silurian echinoids *Palaeodiscus* and *Echinocystites* have lanterns that are similar in all respects except in tooth structure. The pyramids are large and upright with well developed lateral wings that are ridged for the attachment of interpyramidal muscle, just as in modern lanterns. The two halves are firmly united along almost their entire length so that there is no real foramen magnum. Rotulae and epiphyses are fairly standard in structure and compasses are preserved in *Palaeodiscus*. Tooth structure is strikingly different in these two echinoids. *Echinocystites* has broad teeth that are almost flat in cross section. Each is composed of a double series of rather squarish rods that alternate along the mid-line forming a V-shaped pattern (Fig. 3.28). Similar oligolamellar teeth have been described from the Devonian by Jesionek-Szymanska (1979). The teeth of *Palaeodiscus* are weakly crescentic in cross section and more modern in appearance. Abaxially there is a median ridge and two lateral zones. No plates can be

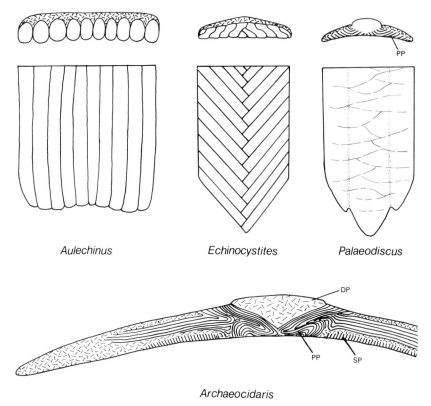

Aulechinus   *Echinocystites*   *Palaeodiscus*

Archaeocidaris

Figure 3.28 Structure of Palaeozoic echinoid teeth. The upper diagram gives a schematic cross section; the lower diagram shows the adaxial face (not given for *Archaeocidaris*). PP, primary plate; SP, side plate; DP area of dense structureless calcite.

seen abaxially and so there is presumably a thin stereom crust covering this surface, as in later teeth. The axial face is weakly concave and shows a double series of inclined tooth plates. These are probably primary tooth plates.

Little further evolution of the lantern took place in the Palaeozoic. Carboniferous archaeocidarids have the typically broad lantern with rotulae and compasses, and pyramids usually have a shallow U-shaped foramen magnum. The teeth are unusually broad and, in cross section, gently curved. As in *Palaeodiscus*, there is an abaxial ridge which appears to be a wedge of dense stereom. The rest of the tooth is made up of a double series of inclined and somewhat folded primary tooth plates with a small adaxial area of secondary tooth plates (Fig. 3.28). The entire adaxial surface is covered in a thin layer of very coarse labyrinthic stereom and a finer stereom covers much of the abaxial surface. No prismatic stereom is present.

In summary, the lanterns of Palaeozoic echinoids were broad and reclined compared to modern lanterns. The presence of compasses, which function as volume compensators, suggests that the lantern in Silurian and later echinoids was able to move in and out of the peristome. This movement could not have been particularly strong, however, as the protractor/ retractor muscles had no specialised perignathic girdle for attachment. The teeth were at first simple, lacking both secondary plates and prism zone stereom, and could not have withstood much bending stress. Teeth must have been used for scooping or biting rather than for scraping, and the earliest echinoids probably fed on surface detritus. Simple secondary plates are present in Carboniferous archaeocidarids and the more robust tooth structure probably allowed them to feed on a wider diet.

The post-Palaeozoic adaptive radiation of echinoids brought about rapid and dramatic changes in the lantern. Just when the prism zone of the tooth evolved is uncertain, but it may have taken place in one of the Carboniferous archaeocidarids. Certainly all post-Palaeozoic echinoids have a zone of prismatic stereom in their teeth. Only the miocidarids survived the Permo-Triassic life crisis and from this group two fundamentally different lanterns evolved – the cidarid type and the euechinoid type (Fig. 3.29).

Cidarid lanterns are narrow and upright and have no foramen magnum. Epiphyses have a deep articulation socket (glenoid cavity) for seating the rotula, and teeth are U-shaped in cross section with a prominently expanded prism zone (Fig. 3.30). For the first time, protractor/retractor muscles attach to a perignathic girdle. The earliest perignathic girdle, seen in the Permian *Miocidaris keyserlingi*, consists of small internal swellings on the most adoral interambulacral plates. These processes, termed **apophyses**, become quite large in Mesozoic and Tertiary cidarids (Fig. 3.32). The perignathic girdle increased the power and efficiency of the retractor and postural muscles. Cidarids have thus evolved a stronger, more active lantern that is able to move in and out with an efficient plucking action and with some degree of lateral movement.

Euechinoids also evolved a perignathic girdle to improve lantern muscle efficiency but, unlike cidarids, it was composed of ambulacral processes (**auricles**) as well as interambulacral processes (Fig. 3.26). This arrangement had the advantage of spreading the muscle attachment areas more evenly around the edge of the corona and so improving the lateral mobility of the lantern (Kier 1974). The euechinoid lantern is relatively narrow and upright in all but the echinothurioids and there is a deep V-shaped foramen magnum making the whole structure lighter. Teeth of primitive euechinoids, such as *Diademopsis*, are crescentic in cross section and have narrow areas of prism zone stereom (Fig. 3.30). This type of tooth is not particularly strong, though obviously an improvement on any of the Palaeozoic teeth. Similar teeth are still found today in the deep-sea echinothurioids, a group that does not use its lantern for rasping.

Selection for stronger teeth, allowing regular echinoids to feed on encrusting organisms, must have taken place early in the Mesozoic, as all

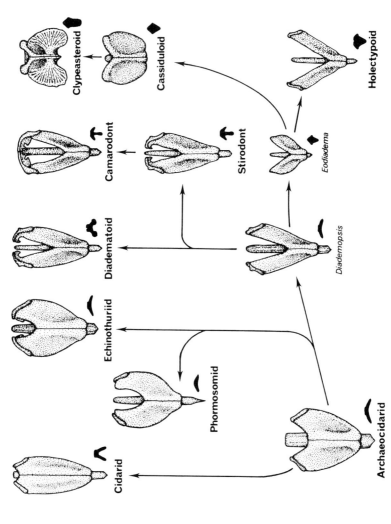

Figure 3.29 Evolution of the lantern in post-Palaeozoic echinoids. Each diagram shows an abaxial view of one pyramid together with a cross section of the tooth in silhouette (from Smith 1981, courtesy of the Palaeontological Society).

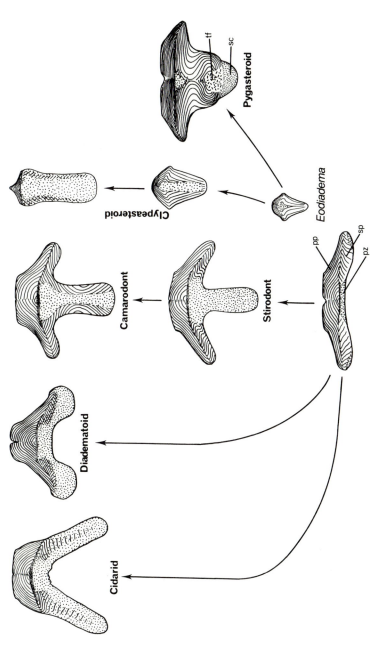

Figure 3.30 Evolution of tooth structure in post-Palaeozoic echinoids. The primitive tooth structure from which other teeth can be derived is based on *Diademopsis*, an early Jurassic pedinoid (see also Fig. 3.10) (modified from Smith 1981, courtesy of the Palaeontological Association).

Key pp primary plate   pz prism zone   sc secondary calcite crust   sp side plate   tf terminal fibres other regular echinoids alive today have mechanically stronger teeth. The diadematoids developed more U-shaped teeth by expanding the lateral prism zones (Fig. 3.30). The teeth of stirodont echinaceans, on the other hand, became stronger by the axial expansion of the prism zone to form a keel (Fig. 3.30). Later, in camarodont echinaceans, the lantern was braced and strengthened further by the expansion of epiphyses which became sutured firmly together above the foramen magnum (Fig. 3.29). This group is today the most successful of all the regular echinoids thanks to its mechanically efficient and highly evolved lantern.

Interestingly, the trace fossil *Gnathichnus pentax* Bromley (Fig. 3.31), which is formed by the rasping action of echinoid teeth, is found only in the Mesozoic and Tertiary. The earliest report of this trace comes from the Rhaetic (Michalik 1977). Palaeozoic echinoids were probably not 'raspers', and it was only the evolution of a more mobile lantern together with mechanically stronger teeth that enabled echinoids to feed in this way.

One group in the Jurassic used their teeth not to rasp, but to pick up and possibly crush edible material. Their teeth were not required to withstand high bending stresses but were subjected more to compressive stress. In these stem irregular echinoids, the tooth evolved greatly expanded side plates enclosing a reduced prism zone (Fig. 3.30). These teeth are diamond- or wedge-shaped in cross section. Primitive irregular echinoids, such as *Eodiadema*, pygasteroids and early holectypoids, retained a functioning lantern throughout life, but in the later groups tube feet took over

Figure 3.31 *Gnathichnus pentax*, the grazing trace produced by regular echinoid teeth. This piece of terebratulid brachiopod shell from the early Bajocian (Middle Jurassic) of the Cotswolds, England has been methodically rasped by a grazing echinoid, presumably feeding on an algal film (× 16).

the role of food collecting, and the lantern was lost or retained only in juveniles.

In the early Tertiary, the lantern was 'rediscovered' by the clypeasteroids. Clypeasteroids probably evolved neotenously from cassiduloids, and the juvenile lantern that they inherited became highly modified as a crushing apparatus. It differs from previous lanterns in several respects. First of all the lantern is wholly internal – not even the teeth are protruded through the small peristome. The muscles running from the pyramids to the perignathic girdle are no longer used for protraction and retraction, but act as jaw openers and postural muscles. Because the lantern as a whole no longer needs to be able to move laterally, the muscle attachment area on the perignathic girdle is small and the modified protractor muscles encircle the retractor muscles (Fig. 3.32). The perignathic girdle is itself highly modified and consists of small ambulacral or interambulacral processes or, in primitive forms, a mixture of both (Kier 1970, 1982). As movement of the lantern is confined to the horizontal plane, it has become inclined and flattened and the teeth have become thick and wedge-shaped (Figs 3.29 & 30). The lantern, if it is used at all, can only act to crush and pulp the material passed into the mouth by the spines and tube feet.

An active lantern raises certain problems, particularly in relation to internal volume control. Movement of the lantern in and out of the test during feeding creates severe changes in internal volume that must be

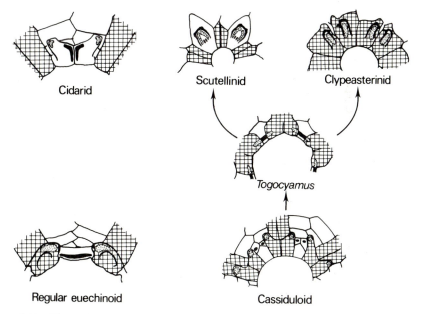

Figure 3.32 The perignathic girdle and muscle attachment in post-Palaeozoic echinoids. Ambulacra cross-hatched; retractor muscle attachment area stippled; protractor muscle attachment area in black (taken from Kier 1970, 1982, courtesy of the Geological Society of America and the Palaeontological Association).

compensated for. By raising and lowering the compasses, the volume of the peripharyngeal coelom is changed and this compensates for movement of the lantern. So the presence of compasses in Silurian echinoids suggests that volume control had already become a problem, and that the lantern was actively protruded and retracted. Cidarids and echinothurioids have large internal pouches connected to the peripharyngeal coelom that act as fluid reservoirs. These are the Stewart's organs, and it is very likely that similar pouches were present in many Palaeozoic echinoids. Acroechinoids have a rather better system in that the pouches to the peripharyngeal coelom are external and can act as expansion sacs. These so-called 'gills' (the term 'gill' is a misnomer since they have little to do with gaseous exchange) extend through notches at the edge of the peristome, and so the presence of buccal slits in fossils is a clear indication that the echinoid had an active, protrusible lantern. Echinothurioids have both expansion sacs and Stewart's organs, and buccal slits first appear in basal Jurassic acroechinoids. Clypeasteroids have an active lantern but, as it is a wholly internal apparatus that is never protruded, problems of volume compensation never arise. Clypeasteroids therefore lack compensation sacs and buccal slits, and have also lost their compasses and peripharyngeal coelom.

Tube feet and spines   Changes in peristomial plating, noted previously, can be correlated with changes in feeding strategy. Palaeozoic echinoids, cidarids and echinothurioids have series of ambulacral plates and tube feet arranged over the peristomial membrane. Peristomial tube feet are small and sensory and lack a terminal disc (Smith 1979). With the evolution, in acroechinoids, of a lantern adapted for plucking and rasping, it became important to have a ring of highly sensitive 'tasters' surrounding the mouth to control operations. The ten buccal tube feet fulfil this need. Each ends in a broad disc richly supplied with nervous tissue (this is reflected in the ambulacral pore structure by the presence of a relatively large notch for the nerve fibres in one of the holes). The disc is supported by a modified rosette which, unlike the rosettes of suckered tube feet, is bilaterally symmetrical and composed of two large and two small elements. The tube foot discs may be so large that they more or less touch, forming an almost complete sensory ring around the mouth.

Echinoids that no longer use the lantern for either rasping or scooping up bottom material, gather their food using tube feet and spines. As food particles are collected by the tube feet, specialised 'tasting' tube feet around the mouth are no longer required and have been lost in irregular echinoids. Several different food-gathering techniques have evolved, and the tube feet and spines may be highly modified for this purpose. We can follow this evolution in fossils from the structure and arrangement of ambulacral pores and tubercles.

In primitive irregular echinoids, such as pygasteroids and holectypoids, there is no tendency for tube feet to become concentrated around the mouth and their tube feet and spines were unspecialised. Ambulacral pores (Fig. 3.33) are like those of the regular echinoids and presumably bore suckered tube feet. These animals were clearly sediment swallowers, as the periproct is displaced from within the apical system, but their method of food gathering could not have been particularly efficient. They probably used both their lantern and tube feet to collect suitable particles. *Echinoneus*, the only living holectypoid to have been studied (Rose 1978), is rather advanced in that it has a lantern only as a juvenile (it is resorbed during growth). In feeding, *Echinoneus* uses its suckered tube feet to collect particles which are then manoeuvred into the mouth with the help of spines.

The first major improvement in food-gathering technique came with the evolution of **phyllodes**. Phyllodes are broadened ambulacral areas of more dense pore (and tube foot) concentration adjacent to the peristome. They first appear in the galeropygoids (Fig. 3.33) and are present in all later cassiduloids and atelostomates. Ambulacral pores in the phyllodes of galeropygoids and cassiduloids are much smaller than the pores of pygasteroids or Jurassic holectypoids and their tube feet must have been correspondingly smaller. These tube feet almost certainly still ended in a suckered disc. As particles smaller than the disc cannot be held by suction, the size of the suckered disc probably limits the size of particles that can be picked up by the tube feet. The smaller size of the tube feet allowed them to pick up smaller particles while the increased number of tube feet around the peristome made them more efficient sediment processors. The increased volume of sediment that could be gathered probably allowed these echinoids to live on less organically rich sediments. However, they never succeeded in exploiting fine substrata because of their suckered tube feet and, today, cassiduloids are confined to coarser sands or gravels.

With the evolution of phyllodes, the spines around the peristome also became modified to help push material into the mouth. These spines form a dense grille across the peristome and attach to the interambulacral plates adjacent to the mouth. Tuberculation here is dense, and individual tubercles are generally radially symmetrical. In cassiduloids, the interambulacral areas may be swollen adorally to form distinct **bourrelets**, increasing the density and number of spines around the mouth. There may also be ten larger tube feet immediately adjacent to the peristome that help to push material into the mouth.

In primitive disasteroids such as *Pygorhytis*, the phyllodes and ambulacral pores are similar to those of cassiduloids (Fig. 3.33) and both groups probably fed in much the same way. However, by the Upper Jurassic the number of ambulacral pores in phyllodes started to decrease and, in *Collyrites* for example, they had also become enlarged (Fig. 3.33). These changes most probably mark the development of specialised tube feet around the mouth that use mucous adhesion, not suction, to pick up particles. Such tube feet are found in almost all living atelostomates and

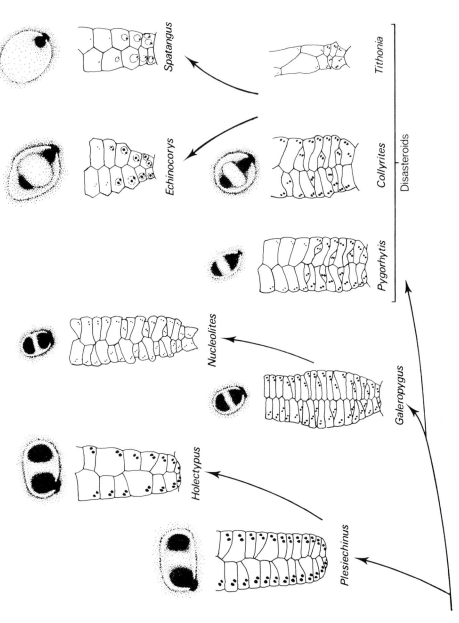

Figure 3.33 Evolution of phyllodes in irregular echinoids. In each diagram the most adoral portion of one ambulacrum is shown with the peristome at the base. Above the plating diagrams one ambulacral pore is illustrated (most plating diagrams taken from Jesionek-Szymanska 1963, courtesy of the author).

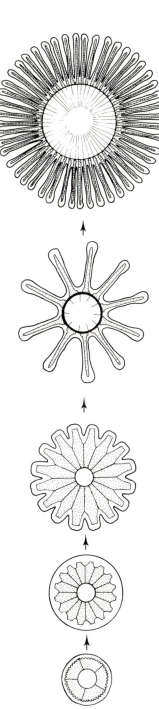

Figure 3.34 Discs of tube feet. A morphological series of tube foot discs showing how a penicillate disc (top diagram) may have evolved from the standard suckered disc of regular echinoids (bottom diagram). All the intermediate forms are known in living irregular echinoids. Skeletal elements stippled.

are termed penicillate. Penicillate tube feet have a broad disc covered in small, sticky finger-like extensions (Fig. 3.19). Although they appear to be far removed in structure from the standard tube feet of regular echinoids, a morphological series can be constructed from other tube feet to show how they might have evolved (Fig. 3.34). The associated ambulacral pore is very distinctive as it is large and nearly circular in shape, with one or a pair of small holes and only a very narrow attachment area for stem muscle forming a rim. Penicillate tube feet are large and prehensile and can pick up 'disc-fulls' of fine sediment by mucous adhesion. Larger particles can be handled with equal facility. The evolution of penicillate tube feet allowed disasteroids and their descendents to exploit finer-grained sediments than had previously been possible. The dense grille of spines covering the peristome was no longer required to help manoeuvre particles into the mouth and was lost. In spatangoids, penicillate tube feet pass their load directly into the oesophagus.

Mucous adhesive tube feet may not be unique to atelostomates. In the Upper Palaeozoic, proterocidarids developed enormously broadened oral ambulacra with a mixture of normal ambulacral pores and large circular pores resembling those of penicillate tube feet (Fig. 3.35). It seems likely

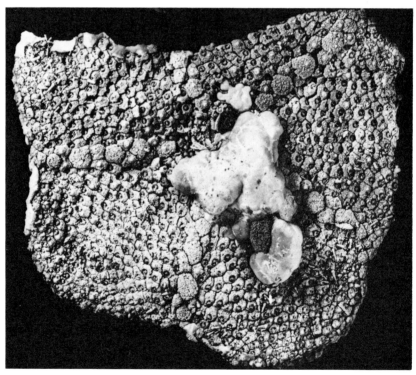

Figure 3.35 The oral surface of *Proterocidaris belli* (Kier) from the Upper Carboniferous of Texas. The ambulacra are extremely broad, and large, round ambulacral pores can be seen (× 0.55) (photograph courtesy of Porter Kier).

that the smaller ambulacral pores were associated with sensory tube feet, whereas the large pores bore tube feet with broad sticky discs that were used to collect sediment.

In certain cases, penicillate tube feet have been abandoned in favour of a more direct method of sediment shovelling. The peculiar bottle-shaped pourtalesiids have only tiny sensory tube feet in their frontal ambulacrum and around the mouth. There is a funnel-shaped area leading to the peristome and its walls are lined with flattened, geniculate spines that probably help to scoop the surface detritus layer into the mouth. Comparable adaptations are found in the Upper Cretaceous holasteroid *Hagenowia* (see Ch. 5).

Spatangoids living deeply buried below the surface have the problem that the nutritional content of the sediment usually decreases away from the surface. Some groups quickly learned to exploit the more organic-rich surface layer by passing material down the anterior ambulacrum. Material is transported in ciliary currents or by spines and mucus. Spatangoids that build a respiratory shaft to the surface may also cascade surface detritus into their burrows using their funnel-building tube feet.

Morphological adaptations associated with this feeding strategy include the development of a sunken frontal ambulacrum with an associated arch of protective spines, and modification of the spines and tubercles within the frontal groove. Sunken anterior ambulacra first appear in Upper Jurassic disasteroids (Kier 1974), become quite common in Cretaceous holasteroids and spatangoids, and reach their greatest development in Tertiary schizasterids. Sinking the frontal ambulacrum creates a channel for transporting material adorally. In order to retain particles within this channel and prevent less nutritious material from the burrow walls clogging or contaminating the flow, a protective arch of spines usually guards the groove. These spines attach to the inwardly facing interambulacral zones immediately adjacent to the ambulacrum, and tubercle arrangement can be used to determine whether such an arch was present in fossil species (Smith 1980b). Within the groove, ambulacral spines are variable though usually small and richly ciliated. Where material is transported in ciliary currents, spines and tubercles in the frontal groove are irregularly arranged, but those feeding using a mucous string have a far more dense arrangement of spines and tubercles that become progressively smaller perradially (Fig. 3.36c). Spatangoids living within coarse sediments or on the surface do not use this strategy to supplement their diet, and consequently lack the morphological adaptations found in sand or mud dwellers (Fig. 3.36a).

The most recent and most spectacular innovations in feeding technique were made during the adaptive radiation of the clypeasteroids that took place in the early Tertiary. Clypeasteroids appear to have evolved neotenously from cassiduloids in the Palaeocene, and are unique in having large numbers of minute tube feet on each ambulacral plate. This innovation allowed clypeasteroids to feed on fine sands, previously unexploited by

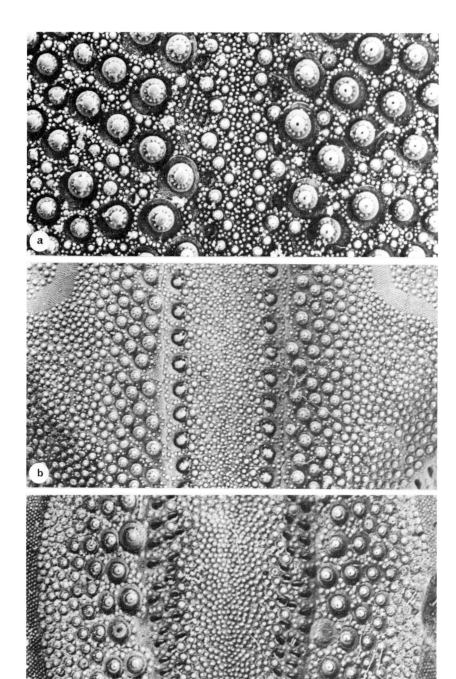

cassiduloids or holectypoids. *Togocyamus*, the most primitive clypeasteroid known, has only an adradial line of pores with relatively few tube
feet per plate (Kier 1982). In later clypeasteroids, tube feet are more
numerous and often arranged in bands. These early clypeasteroids were
very small and material could be passed to the mouth with very little
difficulty as no tube foot was far from the peristome. Fibulariids still feed in
this way today. However, as individual species evolved to become larger it
became relatively inefficient to pass grains adorally from tube foot to tube
foot. Food-gathering efficiency was improved with the evolution of **food
grooves**, ciliated spine-free tracts on the oral surface leading to the peristome. Material can be transported to the mouth along food grooves by
ciliary currents or incorporated into mucous strings or boluses. Tube feet
can then simply place selected particles into the nearest part of the food
groove with the result that more tube feet are free to capture food particles
and a greater amount of material can be dealt with at the same time. Food
grooves are absent in the fibulariids and appear to have evolved twice
independently, once in the Clypeasterina and once in the Scutellina.

In clypeasterinids, the food grooves are straight and simple and lie at the
mid-line of each ambulacrum. Grooves are totally free of tube feet and
their pores and there is no well developed arch of spines (Fig. 3.37a). In
scutellinids the food grooves were also simple at first, an arrangement still
found in the Laganidae. Here, however, the grooves are lined with tube
feet (Fig. 3.37b). Furthermore, branched food grooves have evolved in at

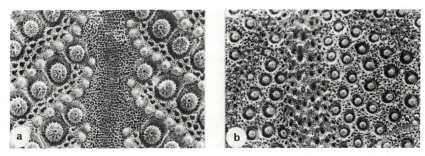

Figure 3.37 Clypeasteroid food grooves: (a) *Fellaster zealandiae* (Recent), a clypeasterinid
sand-dollar (× 40); (b) *Echinarachnius parma* (Recent), a scutellinid sand-dollar. In both
cases the food groove is approximately central and runs vertically with the peristome towards
the top (× 20).

Figure 3.36 The anterior ambulacrum in Recent spatangoids. (a) *Brissus unicolor*: the
ambulacrum is narrow and flush with the surface and is not involved in food gathering and
transportation (× 7). (b) *Brissopsis atlantica*: the ambulacrum is broader and depressed and
has large ambulacral pores for funnel-building tube feet. Ambulacral tubercles are small and
dense but irregularly arranged. Ciliary currents draw fine particulate matter down the groove
(× 10). (c) *Echinocardium cordatum*: the ambulacrum is broad and depressed with many
large ambulacral pores for funnel-building tube feet. Tubercles are arranged in decreasing size
towards the mid-line and food is transported down the ambulacrum in a mucous string (× 10).

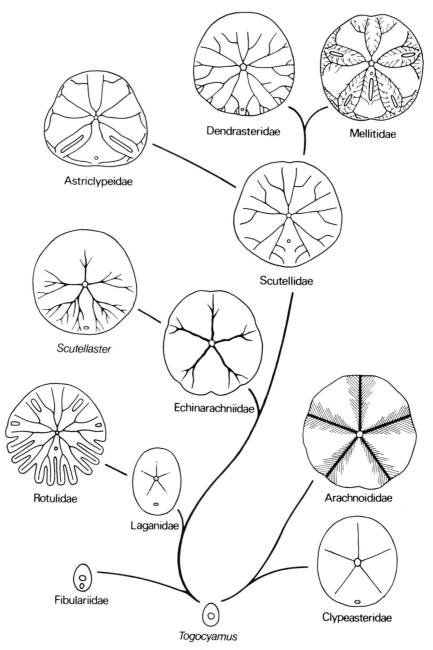

Figure 3.38 The evolution of the food-groove system in clypeasteroids. A branched system of food grooves on the oral surface has evolved several times independently in clypeasteroids.

least three independent lineages (Fig. 3.38). The food grooves of echinarachniids split rather irregularly distal to a long primary trunk, whereas in rotulids and in scutellids, the five primary food grooves bifurcate close to the peristome, either on or at the edge of basicoronal plates. In some groups the food groove system is extensively branched to form a complex network. One advantage of a branched food-groove system is that more tube feet are brought within range of a food groove, thus increasing the bulk of sediment that can be processed. This may have been an adaptation for living in sediments of rather lower organic content.

A second advantage of branched food grooves is that it enables the food-groove system to expand to cover the entire periphery of the test. Sand-dollars have evolved a sophisticated method of sieving out the fine detrital material from the surface layer of sediment, using their aboral spine canopy (see p. 55). This fine material is transported centrifugally on the dorsal surface towards the periphery and around to the oral surface in ciliary currents near the test surface. Because food grooves are branched to cover the entire periphery, the fine material can be captured from all around the margin of the test, so increasing the efficiency of this method of food harvesting. In the most advanced groups, the actual periphery of the test has been lengthened with the development of marginal notches or **lunules** (fully enclosed perforations that develop from notches during growth). This adaptation again appears to have evolved in several groups independently.

In sand-dollar evolution there was also a progressive modification and specialisation of the spines. Around lunules and notches there are large, flattened spines that can control the flow of large particles passing through (Ghiold 1979). Lunules are sometimes lined with cilia-rich miliary spines that can create strong water currents. Spines on the oral surface are arranged to move material from the periphery and the lunules, towards the food grooves where the tube feet can capture them. The associated tubercles show a corresponding variation in size and direction of areole enlargement.

3.3.5 The evolution of feeding strategies

Primitive echinoids from the Upper Ordovician were probably detritus feeders, using their simple lantern to scoop up the surface layer of sediment. The lanterns of later Palaeozoic echinoids had all the elements found in the lanterns of living regular echinoids, although the teeth were very much simpler in structure and mechanically very much weaker. Palaeozoic echinoids were not 'raspers' but must have been rather generalised feeders with a mixed diet of plant material, sessile soft-bodied organisms and detritus. One group, the proterocidarids, specialised as microphagous sediment eaters in the Carboniferous and Permian and evolved large, oral, food-gathering tube feet.

Only miocidarids survived the Permo-Triassic life crisis and all post-Palaeozoic groups arose from them. Mesozoic regular echinoids developed mechanically stronger teeth and lanterns, culminating in the camarodont lantern, and this allowed them to rasp and so feed on encrusting organisms. One early group changed to a diet of sediment, and quickly diversified to give rise to irregular echinoids. Suitable sediment was collected first of all using both the lantern and suckered tube feet, then later using just the tube feet. Later improvements in food-gathering techniques came with the evolution of mucous-adhesive tube feet, that allowed atelostomates to feed on fine-grained sediments, and the development of a frontal groove for channelling material adorally, either suspended in currents or bound in a mucous string.

In the Tertiary, clypeasteroids made the adaptive breakthrough to feeding on fine sands with the evolution of large numbers of tiny tube feet, and rapidly diversified. Fibulariids remained unspecialised but, in both the Clypeasterina and Scutellina, food grooves evolved to increase the efficiency of sediment handling. By the Eocene, sand-dollars had evolved a most efficient food-gathering technique whereby the surface layer of sediment was sieved for its detritus. This technique was improved with the development of a branched food-groove system and the extension of the periphery by lunules or notches. These innovations have allowed sand-dollars to diversify and thrive in sediments of low nutritional value and they have remained successful to this day.

3.4   Reproduction and growth

The reproductive cycle in echinoids and their larval and post-larval development have been widely studied, though largely with reference to soft tissue structures. In this chapter only those aspects of reproduction and growth that affect skeletal structures, and which can therefore be retrieved when working with fossil material, will be examined. For a more biological account the most authoritative descriptions of skeletal development and growth are still those of Gordon (1926, 1929), while good reviews of reproduction and growth can be found in Boolootian (1966).

3.4.1   Reproduction

Recent regular echinoids have five gonads suspended from the interior of the test in an interradial position. The sac-like gonads taper adapically and each opens to the exterior through the gonopore that perforates each genital plate. The five gonads are not independent but are linked to one another by a connecting ring that lies just internal to the genital plates.

Except for the rare freak, the sexes are always separate in echinoids. The two sexes are morphologically indistinguishable in the great majority of species, though sexual dimorphism does exist and can be recognised in fossils. Males and females occur in roughly equal numbers except in species where the offspring are brooded by the mother, when females greatly outnumber males. For example, populations of the brooding irregular echinoid *Cassidulus cariboearum* have five times as many females as males (Gladfelter 1978).

Individuals usually become sexually mature towards the end of their first year 'though short-lived species that grow rapidly may mature within months of settlement, and longer-lived species may not become sexually mature until their second or even third year. The size at sexual maturity is quite variable, even between members of the same species. Table 3.1, (p. 88) which lists the age and size at maturity and the maximum age and size attained for a range of echinoids, gives some indication of the variation that can be encountered.

Gonadal tissue takes approximately six months to mature (Pearse & Phillips 1968). When ripe, the gametes are shed into the water and fertilisation takes place externally. Spawning is synchronised in temperate and cold-water species, but may be asynchronous in some tropical and subtropical species. The controls that govern the timing of spawning are not yet clear, but changes in temperature or the onset of plankton blooms have been implicated (Himmelman 1978). Pearse (1969) concluded that gonad maturation will only begin once the urchin has built up a critical minimum food reserve, and water temperature has reached a critical minimum level. However, Thompson (1983) has recently demonstrated that a reduction in the ration of food will also trigger off an increase in the rate of gonadal growth. It has been reported that, in some species, individuals aggregate prior to spawning, but it has yet to be shown that this is the primary reason for herd formation. Close synchronisation of spawning can be achieved within populations since the presence of gametes in the water acts as a stimulus to trigger off spawning in other ripe individuals situated downstream.

In the majority of species the gametes are small, eggs and sperm are similar in size and the adults show no sexual dimorphism. Following fertilisation, a free-swimming pluteus larva develops. However, there are species in which the females produce a small number of large, yolk-rich eggs which are retained amongst the spines or within special depressed zones on the test. According to Gladfelter (1978) fertilisation may take place internally. The fertilised eggs develop directly without passing through a free-swimming larval stage. There are usually secondary sexual characters in adults of these species. Brooding has developed independently in several lineages and has been recorded in cidarids, echinaceans, cassiduloids, clypeasteroids, holasteroids and spatangoids. Most brooding species are cold-water forms, but there are also tropical brooding species such as *Cassidulus*. Brooding is presumably adopted where a wide dispersal of larvae would be disadvantageous as, for example, where the species is restricted to a very specific habitat or where predation of larvae is very heavy.

The information on reproduction that can be recovered from fossils is fairly limited and concerns the internal arrangement of gonads, the size at sexual maturity and the gender in dimorphic species.

Gonadal arrangement   The number of gonads can quickly and easily be assessed by counting the gonopores. Although all Mesozoic and Tertiary regular echinoids had five gonopores and thus five gonads, Palaeozoic echinoids were quite different. Ordovician and Silurian echinoids possess only a single genital/madreporite plate and presumably had just a single gonad lying in that interray. Embryology provides some support for this interpretation since, in living echinoids, gonadal tissue originates in the interray with the madreporite and only later grows around to form a ring of tissue beneath the other genital plates. From the Devonian onwards, Palaeozoic echinoids, whose apical systems are known, all have five genital plates, one of which is usually enlarged as a madreporite. Unlike post-Palaeozoic echinoids, each genital plate is perforated by between 3 and 13 gonopores. Presumably a ring of gonadal tissue must have existed beneath the genital plates but this, instead of giving rise to just one large gonad in each interambulacrum, probably had a series of small gonads hanging down from it. The oldest echinoid that I have seen with genital plates perforated by a single gonopore is the Triassic miocidarid *Lenticidaris*.

During the evolution of irregular echinoids, the posterior gonad was quickly lost as the periproct migrated backwards out of the apical system. The Lower Liassic pygasteroid '*Plesiechinus*' *hawkinsi* has five gonopores (Fig. 3.22) but almost all other Mesozoic irregular echinoids have lost the fifth gonopore. The number of gonads decreases still further in some Cretaceous and Tertiary atelostomates where there are only two or three gonopores, but the reasons for this are not always clear. Within the holectypoids and clypeasteroids several groups have independently re-evolved a fifth gonad and gonopore.

Sexual maturity   Gonopores are formed by skeletal resorption (Gordon 1926). As gonopores do not appear until the gonads have begun to mature (on reaching a size of between 2 and 40 mm diameter depending upon the species) their time of appearance marks the onset of sexual maturity. This provides a simple way of distinguishing between juveniles and adults of fossil species.

Sexual dimorphism   Sexual dimorphism is found in species that brood their young. The morphological differences which can be used to determine gender in fossils have been reviewed by Kier (1969) and Philip and

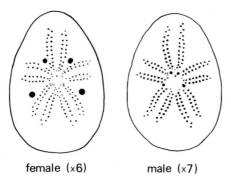

female (×6)   male (×7)

Figure 3.39 Sexual dimorphism in the Eocene clypeasteroid *Echinocyamus bisexus* (from Kier 1968). The female has larger and more widely separated gonopores than the male.

Foster (1971). The sexes can be distinguished on the basis of the following features:

(a) Gonopores are larger and placed further apart in females than males (Fig. 3.39). This is because females produce large yolk-rich eggs. Müller (1970) has identified this type of sexual dimorphism in the Cretaceous cidarid *Stereocidaris*.

(b) Females in some (though not all) brooding species have depressed zones on the test in which the young are brooded. These pouches vary both in shape and position amongst the 45 or so species known to possess them (Fig. 3.40). The entire apical system may be sunken or there may be an annular depression adapically. Two species of cidarid have a circum-oral brood pouch and in some clypeasteroids there are an anterior pair of oral interambulacral pouches. The temnopleurid *Pentechinus* has five deep adapical interambulacral pouches. Finally, the spatangoid *Abatus* uses its deeply sunken petals as brood pouches, and Lambert (1933) has described and illustrated a Cretaceous specimen of *Abatus* with juveniles preserved *in situ* within the petals.

(c) Females usually grow considerably larger than males and within a population females significantly outnumber males.

3.4.2   Larvae and larval development

Except in brooding species, the fertilised egg develops into a free-swimming larva known as the echinopluteus. The echinopluteus develops elongate arms so as to extend its ciliary bands and these arms are supported by skeletal rods (Fig. 9.3). The larval skeleton consists of up to three lateral paired rods plus a single anterior and posterior rod, all linked to form a more or less cohesive framework. Fossil spicules attributable to echinoplutei have been described by Kryuchkova and Solov'yev (1975).

The free-swimming larval stage lasts approximately for one or two months, during which time considerable morphological changes take place.

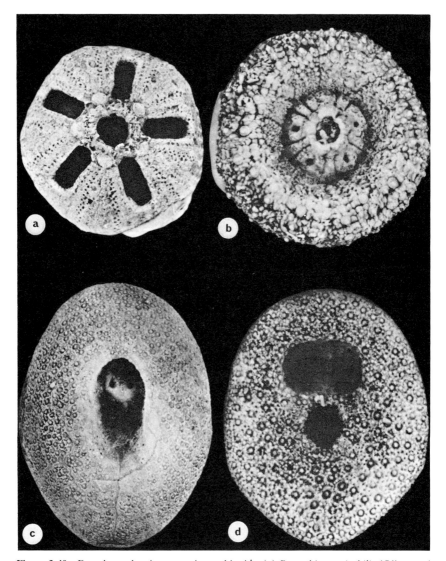

Figure 3.40 Brood pouches in marsupiate echinoids. (a) *Pentechinus mirabilis* (Oligocene): apical view (× 5) showing five large interambulacral pouches. (b) *Paradoxechinus novus* (Oligocene): apical view (× 5). An annular depression surrounds the periproct. (c) *Peraspatangus brevis* (Miocene): apical view (× 5). The entire apical system is deeply sunken. (d) *Fossulaster halli* (Miocene): oral view (× 8) showing large anterior pouch. (All figures from Philip & Foster 1971, courtesy of the Palaeontological Association.)

This stage ends when the pluteus metamorphoses into a juvenile and settles. The newly metamorphosed echinoid is generally around 0.5 to 1.0 mm in diameter and already possesses a complete apical system, some four or more plates in each ambulacrum and interambulacrum, lantern elements and spines.

3.4.3 Post-larval growth

Growth rate is initially rapid in the first year of life but decreases as the echinoid grows older. Test growth is achieved in two ways: (a) by adding new plates at the apical end of ambulacral and interambulacral columns, and (b) by increasing the size of existing plates. As the number of plates forming the test is directly proportional to size rather than age, the rate of plate addition slows down as the echinoid grows older, though plates generally continue to be added throughout life. There are some exceptions to this. Kier (1956) pointed out that, in some spatangoids, interambulacral columns become separated from the apical system in large specimens, showing that interambulacral plate addition had ceased. Similarly, in many laganid clypeasteroids the presence of a single unpaired plate at the apex of each interambulacrum shows that plate addition has ceased in these areas.

As growth proceeds, plates that were initially aboral shift round to the oral surface (Fig. 3.41), ambulacral and interambulacral plates always maintaining their same position relative to one another. In sand-dollars and many heart-urchins there are only a small number of ventral plates which become established very early on in life. During later growth, plates continue to be added adapically but no further plates move adorally around the ambitus (Fig. 3.41) and adoral plates become very large to compensate.

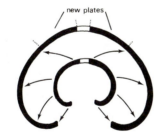

Regular echinoid

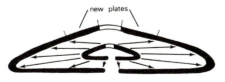

Sand–dollar

Figure 3.41 Diagrammatic representation of the expansive growth of the echinoid test. Test growth occurs by enlarging pre-existing plates and by adding new plates apically. In the regular echinoids plates move adorally around the ambitus as growth proceeds but in sand-dollars the number of oral plates becomes fixed at an early stage. The peristome enlarges through plate expansion and not by resorption.

The test of regular echinoids is globular so ambital plates are always larger than older plates lying adorally. However, this does not imply that plate resorption dominates below the ambitus. This was elegantly demonstrated by Deutler (1926) who used growth lines within plates to show that resorption did not occur, and that differences in plate size were produced by differences in relative growth rate. Even the widely held belief that the first-formed interambulacral plates are totally resorbed at the peristome in the regular echinoids has been disproved in at least two species (Märkel 1981).

Growth rates are variable, even within a single individual, (for example the sand-dollar *Mellita quinquiesperforata* grows between 0.5 and 9.5 mm in diameter per month – Lane & Lawrence 1980). Growth rate is affected by a variety of different factors including temperature, reproductive state, availability and type of food, and weather (stormy weather inhibits feeding and increases the injury rate). Seasonal variation in these factors imparts an annual cyclicity in growth rate, which is particularly marked in temper-

Table 3.1 Size and age at sexual maturity for selected echinoids.

	Maximum test diameter and approximate age		Size and age at sexual maturity	
	mm	years	mm	years
CIDAROIDS				
Eucidaris tribuloides	53	?	20	1
DIADEMATOIDS				
Diadema antillarum	76	2–3	30	< 1
ECHINACEANS				
Allocentrotus fragilis	70	7.5	35	2–3
Echinometra mathaei	65	2	20	≪ 1
Evechinus chloroticus	145	10–15	40	2–3
Loxechinus albus	106	5	20	< 1
Lytechinus variegatus	79	2–3	30–40	< 1
Psammechinus miliaris	62	12	5–7	1
Tripneustes ventricosus	110	1–2	30	≪ 1
CASSIDULOIDS				
Cassidulus cariboearum	35	2–3	9♂, 15♀	1
CLYPEASTEROIDS				
Arachnoides placenta	36	> 5	18	1
Dendraster excentricus	80	9	12–20	1
Pentedium curator	3.3♂, 4.5♀	–	2–2.7	–
SPATANGOIDS				
Echinocardium cordatum	50	> 10	20–25	1–2
Meoma ventricosa	165	4	20–40	< 1
Schizaster floridiensis	16	–	6	–

ate and cold-water species. When conditions become harsh, growth ceases and the test may even shrink in size through resorption (Ebert 1967, Lane & Lawrence 1980). The life expectancy for echinoids ranges from less than a year to approximately 15 years (Table 3.1) but these data come almost entirely from shallow-water species and little is known about the longevity of deeper-water species.

In a fascinating analysis of growth and mortality data on post-larval echinoids, Ebert (1975, 1982) found that growth and mortality rates are positively correlated in all those species for which data are available. Rapidly growing echinoids are relatively short lived, whereas slower growing echinoids are generally longer lived. Ebert suggested that this correlation arose because echinoids have only limited resources that have to be shared amongst gonadal growth, somatic (body) growth and general maintenance (resources put into maintenance are used to repair and strengthen the test and into anything that increases the probability of individual survival). One obvious effect of this is that somatic growth usually slows or ceases during periods of gonadal growth.

The manner in which resources are allocated produces different life strategies. In temporarily suitable habitats where survival is unpredictable, echinoids generally grow rapidly and reproduce as early as possible. They achieve this by allocating proportionally more of their resources to gonadal and somatic growth and less to maintenance than do slow-growing echinoids. In habitats where competition is high and successful recruitment cannot be guaranteed every time, echinoids channel more of their resources into maintenance at the expense of growth and, in this way, try to survive and spawn over a number of years. The strategy adopted by an echinoid affects the way in which the skeleton grows and is discussed in the next section.

3.4.4 Plate growth

Skeletal growth has been studied in a variety of species and using a variety of techniques, namely growth-line analysis, radioactive labelling and computer modelling. Plates grow by peripheral accretion, with the fastest growth at the perradial or interradial suture and slowest growth at the adradial or adapical suture. Plates usually increase in thickness as they grow, particularly during the early stages. In order to maintain a uniform plate thickness as the plate thickens peripherally, a filler stereom is deposited centrally on the inner surface. In many species a history of plate growth is recorded in the skeleton in the form of growth lines.

Growth lines in plates show up as alternating bands of different stereom in thin section or under a scanning electron microscope (Pearse & Pearse 1975, Smith 1980c). In the middle plate layer, galleried stereom is formed seasonally during periods of growth, whereas more porous labyrinthic stereom forms as growth slows and ceases. In the inner plate layer,

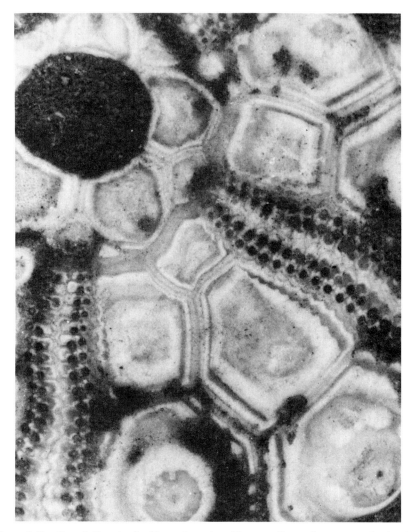

Figure 3.42 Growth lines in the Upper Jurassic echinacean *Hemicidaris intermedia* (× 8). The plate surface has been worn away to reveal light and dark growth banding in apical, ambulacral and interambulacral plates suggesting that this individual is some three years old (photographed under water).

labyrinthic stereom forms during periods of faster growth and denser perforate stereom when growth has more or less stopped. Growth lines are usually only visible in fossils where the plate surface has been worn (Fig. 3.42). They can be studied in thin section or by applying weak acid which picks out differences in stereom porosity. Growth lines are very useful and can be used to determine the age of fossil echinoids, identify repaired structures and interpret life strategies.

Age determination   Growth lines are well developed in species that show a clear seasonality in growth with a cyclic alternation between somatic and gonadal growth. Major growth bands are generally considered to be produced annually, but in detail may consist of a number of closely spaced stereom discontinuities. The presence of growth lines in fossils can be used as a direct measure of the individual's age. However, as plates are constantly being added apically during growth, only ocular and genital plates, together with the more adoral coronal plates, will possess the full number of growth lines. Furthermore, growth lines can be produced by any factor that inhibits somatic growth (notably starvation) and not purely by seasonal factors. Growth line data must therefore be treated with caution.

Regeneration   Skeletal elements can be repaired and regenerated following breakage and spines are especially prone to damage. The solid spines of cidarids and echinaceans have a stereom growth banding just like plates. Spines that have been broken and repaired are easily recognisable in longitudinal section, since the regenerated tip lacks earlier growth lines that are present in the lower part of the spine.

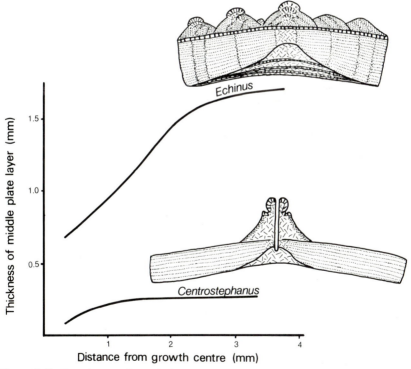

Figure 3.43 Growth strategies and plate structure. Two contrasting strategies are shown. *Centrostephanus* grows rapidly and puts little effort into plate thickening and strengthening, whereas *Echinus* grows proportionally slower and continues to increase the thickness of its plates over several years.

Resource allocation   Ebert's model of resource allocation amongst somatic growth, gonadal growth and maintenance has important implications for
plate growth, and different life strategies produce differences in plate structure. Somatic growth increases test size by peripheral accretion to individual plates. Stereom discontinuities (growth lines) usually mark periods
when resources have been switched from somatic growth to gonadal
growth, and so record annual reproductive cycles. The rate at which plate
thickness increases relative to plate area can be measured in cross section
by plotting the thickness of the middle plate layer against distance from the
centre (Fig. 3.43). This reflects the relative emphasis given to somatic
growth as opposed to maintenance (i.e. plate strengthening). Echinoids
that grow slowly, reproduce over a number of years and put some effort
into body maintenance will typically have plates that show several growth
lines, and which increase in thickness throughout most of their lives.
Opportunistic species that grow extremely rapidly and then spawn with
little allocation to maintenance, have thin plates that grow with little thickening and have few, if any, growth lines. These two extreme life strategies
produce plates quite different in structure (Fig. 3.43) though, of course, all
intermediates exist. The plate structure of fossil echinoids can then be used
to form some idea of what their life strategies were. For example, the
thick-plated Carboniferous palaechinoids were probably slow growing,
long lived echinoids, whereas the contemporary archaeocidarids appear to
have been rapidly growing opportunistic species.

3.4.5   Allometric growth

Structures that grow allometrically usually do so for a good reason. For
example, in regular echinoids the periproct grows proportionally more
slowly than the test in diameter. This presumably balances the necessity, in
juveniles, of having a large enough periproct to cope with proportionally large faecal particles, without reducing the strength of the test too much
in adults. Similarly, lunules grow allometrically faster than body length in
sand-dollars, correlating with the well known fact that an animal's appetite
grows faster than its linear length (Alexander & Ghiold 1980).

    Allometric growth is particularly noticeable in the spines of irregular
echinoids. All spines and tubercles show a certain degree of allometry,
growing rapidly at first but more slowly later compared with test diameter.
In regular echinoids, spines and tubercles generally increase in size throughout life and the spacing between spines progressively increases. However,
where spine density is critical, as for example when the spine canopy must
hold the overlying sediment away from the test surface, tubercle and spine
spacing is fixed throughout life, and spines reach their final size and shape
in one rapid burst of growth. As they undergo no further growth, these
spines can be highly modified and often end in an enlarged, flattened
or bent tip.

3.5 Gaseous exchange

The transport of oxygen to, and carbon dioxide from, internal tissue poses problems to any globular animal because of its small surface area to volume ratio. In addition, echinoids have only a very primitive circulatory system of coelomic fluid that lacks respiratory pigments. Although some gaseous exchange can take place by direct diffusion through the body wall, the tube feet and ampullae provide a much more efficient pathway. The circulation of coelomic fluid between tube foot and ampulla quickly and efficiently transports dissolved gases through the body wall. The gases need only diffuse through the thin walls of the tube foot and ampulla. The tube foot–ampulla system also tremendously increases the surface area available for gaseous exchange. In a 70 g regular echinoid *Strongylocentrotus*, tube feet increase the external surface area by some 2.5×10^4 mm^2 (Steen 1965) and ampullae add a similar area internally. Although all tube feet allow gaseous exchange to take place, specialised tube feet are often developed specifically for gaseous exchange.

The rate of oxygen consumption in echinoids, as in most living systems, is dependent upon temperature. At low temperatures (below 6°C) diffusion through the body wall is sufficient to effect the great majority of gaseous exchange, but at 16°C oxygen consumption is directly proportional to the surface area for gaseous exchange, and the number and efficiency of the tube feet are limiting factors.

As metabolic rate and oxygen consumption are proportional to temperature, echinoids living in warm waters require more efficient gaseous exchange systems than those from colder waters. For example, the holasteroids which now live in cold deep-water habitats lack specialised respiratory tube feet, unlike many of their Mesozoic ancestors that lived in shallower habitats. Similarly, cidarids from deeper and colder habitats lack specialised respiratory tube feet whereas species living in the shallow warm waters associated with reefs all have highly specialised tube feet (Smith 1978b).

3.5.1 Morphological adaptation

Although tube feet are never preserved in fossil echinoids, we can follow the evolution of respiratory specialisation through the morphological changes that took place in ambulacral pores. During the evolution of echinoids, gaseous exchange was improved in the following five ways, each of which affected pore structure in a recognisable way.

(a) *Evolution of a one-way circulation system between tube foot and ampulla*. This is achieved by having a pair of pores connecting the tube foot to its ampulla. Coelomic fluid flows into the tube foot through the adradial pore and out of the tube foot via the perradial pore. Tube feet without any respiratory function are connected to their ampullae by a single pore and coelomic fluid moves in and out of the tube foot only when it extends and retracts.

(b) *Separation of oxygenated and deoxygenated currents within tube feet and ampullae*. It is important that the incoming and outgoing currents are prevented from mixing in the tube foot and ampulla. In the simplest case this is achieved by having a partitioning septum running up much of the length of the tube foot. However, diffusion can still occur across this barrier, reducing the efficiency of the system and so, to prevent this, the two main channels that carry fluid into and out of the tube foot and ampulla are usually separated by a central zone of horizontal partitions (Fig. 3.44). The associated ambulacral pore structure is very distinctive with the two pores widely separated and linked by a narrow groove. The two pores usually diverge as they pass internally, a sure sign that the ampulla is a flattened and leaf-shaped structure specialised for gaseous exchange.

(c) *Reduction in thickness of the tube-foot wall*. Gaseous diffusion is speeded up by reducing the thickness of the barrier to be crossed. Whereas suckered tube feet need strong, muscular stems with lots of muscle and collagen fibres, respiratory tube feet need only a few contractile fibres and a thin connective-tissue layer. Tube feet specialised for gaseous exchange are exceedingly thin-walled and so the associated ambulacral pores either lack, or have an extremely narrow rim of attachment stereom.

(d) *Increase in the relative surface area of tube feet and ampullae*. An easy way to improve the efficiency of gaseous exchange within a system is to increase the surface area to volume ratio. Tube feet and ampullae

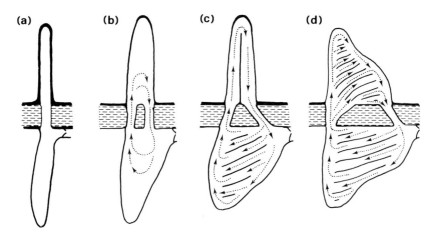

Figure 3.44 Evolutionary improvements of the tube foot/ampulla system for gaseous exchange: (a) the primitive condition; (b) tube foot and ampulla connected by a double pore allowing a one way circulatory system, thinner walled tube foot and ampulla; (c) septum in tube foot and partitions in ampulla to increase circulation pathway and separate inflow from outflow; (d) tube foot flattened (large surface area) and partitioned to separate further inflow from outflow.

involved in gaseous exchange generally become flattened and elongate. Obviously the associated ambulacral pore must also become elongated and, in order to achieve a more uniform current distribution within the tube foot, the adradial pore typically becomes long and narrow and may even become partitioned.

(e) *Positioning of respiratory tube feet on the test.* Tube feet specialised for gaseous exchange always develop aborally where there is better water circulation. This is particularly crucial to infaunal irregular echinoids. Infaunal irregular echinoids rely on their densely ciliated miliary spines to create currents, and draw water down through the sediment and past aboral tube feet. Only heart-urchins have been able to live successfully in less permeable fine-grained sediments, because they can construct and maintain a respiratory shaft to the surface, and have organised bands of ciliated spines (**fascioles**) for creating current.

3.5.1   Evolution of the gaseous exchange system

The earliest echinoids lacked any specialisation for gaseous exchange whatsoever. Ambulacral pores in Ordovician echinoids are either small, single and positioned perradially or, as in *Eothuria*, consist of a circular zone of tiny holes. By the Lower Silurian, a one-way circulation system had developed in tube feet and all ambulacral pores were double. Little further development took place during the Palaeozoic and tube feet and ampullae apparently remained tubular and unspecialised. There was, however, a division of labour amongst tube feet in some groups, with thin-walled suckerless tube feet aborally, and food-gathering or suckered tube feet orally.

It was not until the Jurassic that species with ambulacral pores associated with specialised respiratory tube feet first appeared. Tube feet specialised for gaseous exchange evolved independently within three groups: cidarids, arbaciids and irregular echinoids. Amongst irregular groups, eognathostomates apparently lacked specialised respiratory tube feet, and respiratory tube feet first developed in galeropygoids. Judging from pore structure, the change from simple cylindrical tube feet to flattened leaf-shaped tube feet took place very rapidly in irregular echinoids. Respiratory tube feet have been secondarily lost in a few later groups, as for example in living holasteroids.

3.6   Defence

Echinoids must avoid or deter potential predators and need some defence against pests and parasites. Three broad strategies can be employed to avoid predation: camouflage; having a deterrent armoury of spines and pedicellariae; and active avoidance. This section deals with the morphological adaptations associated with defence.

3.6.1 Camouflage

Some echinoids try to avoid predators by camouflaging themselves. Cidarids, for example, provide many good examples of camouflage using spines. As the shafts of cidarid spines are not covered by soft tissue they commonly become encrusted by algae and epizoans which effectively camouflage the animal. The flattened, fan-shaped spines that are found in several groups of cidarid probably evolved to encourage the settlement of algae and epizoans and enhance the camouflage. A covering of misshapen, often grotesque spines, as in *Aspidocidaris*, breaks up the outline of the animal. Cryptic species of cidarid are all found in reef-associated habitats where this type of camouflage can be very effective.

Covering is another technique used by several groups to help break up the outline of the animal. Surrounding loose material on the sea floor is picked up and held aloft by aboral tube feet so that the echinoid blends in with its surroundings. During the night much of the cover is dropped and fresh material has to be collected at the start of each day. Although various people have suggested that covering might give protection from the harmful effects of ultraviolet radiation, from desiccation or from current surge, an important and probably the prime function of this reaction is to provide camouflage. A large number of echinaceans cover themselves as do some semi-infaunal irregular echinoids such as *Clypeaster*.

Covering is only possible in regular echinoids that have suckered discs to their aboral tube feet, and all the regular echinoids that have aboral suckered tube feet do apparently cover themselves to some extent. It should therefore be possible to identify those fossil species that could have used this cryptic strategy. The ambulacral pores for suckered tube feet differ from pores associated with respiratory or sensory tube feet in that they are surrounded by a narrow but distinct rim of attachment stereom for the insertion of stem muscle fibres (see p. 41). Furthermore, in species that cover, the aboral spines are uniform in length and always relatively short, so that the tube feet can extend well beyond them to manoeuvre material. The primitive irregular echinoid *Plesiechinus* apparently had aboral suckered tube feet and relatively short spines and may have covered itself (Smith 1978a).

3.6.2 Defensive armoury

Spines are undoubtedly the most important weapons of defence against would-be predators. They can be used in one of three ways. First, the spines may be toxic and carry a bulbous poison sack at their tip. Such spines are found in echinothurioids where they are slender, hollow and thin-walled. The large central lumen of the spine opens distally. Many echinoids also carry globiferous pedicellariae which likewise have poison sacks and help to protect them against predators.

Figure 3.45 *Tylocidaris clavigera*, an Upper Cretaceous cidarid with large club-shaped spines (× 2).

A second approach is to have long, slender and needle-sharp spines to deter possible predators. This is the defence used by diadematoids. Their spines are relatively brittle and tend to break, leaving the tip embedded in the predator. Diadematoids increase the effectiveness of this defence by aggregating in large herds on the open sea floor and, when disturbed, clustering their spines towards the intruder. Despite this formidable array of spines, however, a variety of fish are able to snap their way through the spine canopy with apparent ease (see Ch. 2).

The third and most widely used ploy is to have extremely large and heavy spines that cannot easily be broken and which make the echinoid too unpalatable and too awkward to handle. The spines of *Acrosalenia* and *Tylocidaris* (Fig. 3.45) are well fitted for this role.

3.6.3 Active avoidance

Echinoids can avoid visual predators by remaining hidden during daylight hours and only emerging to forage at night. Many regular echinoids have adopted this mode of life and wedge themselves into nooks and crannies during the day. Probably for the same reason, the holectypoid *Echinoneus* lives under loose coral debris during the day and only emerges at night (Rose 1978). Unfortunately, as this is purely a behavioural adaptation there seem to be no morphological criteria for recognising fossil species that avoided predators in this way.

Irregular echinoids are able to avoid a number of predators by living infaunally. Whether they adopted an infaunal mode of life to escape predation is, however, arguable. Reduced predation rate might simply have been an added bonus when irregular echinoids moved into areas of unconsolidated, mobile sediment and had to burrow in order to gain stability in currents.

3.6.4   Pedicellariae

An echinoid's main defence against settling larvae and other small pests are the tiny pincer-like organs termed **pedicellariae** which are scattered amongst the spines. There are four broad types of pedicellariae; tridentate, globiferous, ophicephalous and triphyllous (Fig. 3.46). Tridentate pedicel-

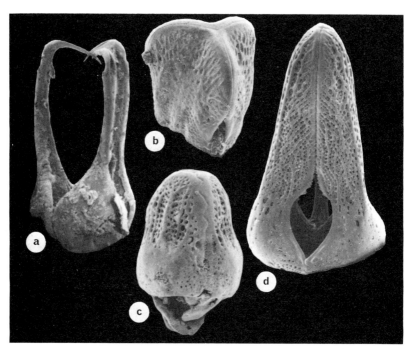

Figure 3.46 Types of pedicellariae from the regular echinoid *Lytechinus variegatus* (Recent): (a) globiferous pedicellaria ($\times$ 36) with needle-sharp jaws for injecting venom; (b) triphyllous pedicellaria ($\times$ 180) with simple jaws for general cleansing activities; (c) ophicephalous pedicellaria ($\times$ 36) with powerful toothed jaws for capturing and holding living organisms; (d) scissor-like tridentate pedicellaria ($\times$ 36) for cleansing activities. All figures are scanning electron micrographs (soft tissue has been removed).

lariae are the largest and simplest type. They consist of a thin calcite-supported stem, a shorter flexible neck and (usually) three long finger-like blades. The blades meet distally, where they are serrated, and expand at the base for attachment of the muscles that open and close the jaws. Globiferous pedicellariae have narrow blades which end in a sharp tooth and carry one or two poison glands. These glands lie internally in pedicellariae of cidarids and open through a pore in the distal tooth, but are external in all euechinoids. Globiferous pedicellariae are usually rather squat, but those of echinothurioids are peculiar in being long and slender like tridactylous pedicellariae. Ophicephalous pedicellariae are the most common type in the regular euechinoids. Their blades are short and broad, and are serrated along almost their entire length. Proximally, there is a very distinctive median ridge on the inner surface. Triphyllous pedicellariae are the smallest and most flexible type. They have simple flat blades that lack serrations.

The functions of these different types of pedicellariae have been investigated by Campbell (1973, 1974, 1976). Globiferous pedicellariae form an important part of the echinoid's defence against predators and pests. They will open with either tactile or chemical stimulation and snap shut when sensory cilia positioned on the inside of each valve are touched, injecting their poison into the unfortunate captive as they grip it. Even when forcibly detached from the test, the pedicellariae will continue pumping in their poison. Ophicephalous pedicellariae are geared to capture and hold active objects and will release inanimate particles. They presumably function in trapping any small organisms that venture in amongst the spines, and those on the peristomial membrane may even help supplement the diet. Tridentate pedicellariae are activated by more continuous mechanical stimuli, and are more suited for picking up inanimate objects. They may function in cleaning the test of debris and deterring pests. Triphyllous pedicellariae are very active and flexible and are constantly on the move. Their main functions are probably to deter larvae from settling on the test and to help clean the test of debris.

Pedicellariae are not uncommon in well preserved fossil material, and information is now available on the pedicellariae present in a variety of fossil species. The oldest known pedicellariae come from Upper Silurian lepidocentrids and echinocystitids (Philip 1963, Blake 1968). These are extremely simple, stemless 'tridactylous' pedicellariae composed of two or three 'blades' attached to a single tubercle. Each 'blade' is circular in cross section and swollen at the base, and both in size and shape is just like any of the adjacent spines. Pedicellariae in the Carboniferous *Archaeocidaris* are much more modern in appearance. *Archaeocidaris* has just two types of pedicellariae, both with calcite stems (Lewis & Emson 1982). The larger type are obviously tridentate pedicellariae, but the smaller type are more difficult to place, since they have short blades like tryphyllous pedicellariae but are not distally flattened. Available data suggest that functional and structural diversification of pedicellariae did not take place until the Mesozoic, and all four types of pedicellariae had certainly evolved by the Jurassic.

Pedicellariae then appear to have evolved from groups of two or three spines connected at their bases to a common tubercle. Their simple form and lack of manoeuvrability suggests that they initially functioned as a deterrent to small marauding pests and that other functions are secondary modifications.

4 Retrieving biological information — a case study

The previous chapter dealt in rather general terms with how echinoids have adapted to perform the various basic life functions, stressing morphological adaptations that can be recognised in fossils. Here I shall try to show how, by using a functional morphological approach, information about the palaeobiology of fossil echinoids can be recovered. As an example I have chosen to examine the Jurassic irregular echinoid *Holectypus*. *Holectypus* has been chosen, first because it is a primitive irregular echinoid morphologically intermediate between regular echinoids and typical extant irregular echinoids, with no directly comparable living representatives. Secondly, *Holectypus* and the almost identical genus *Caenholectypus* were extremely successful and achieved world-wide distribution during the Mesozoic, so that most teaching collections in Europe and America should have at least some specimens. Although this account refers specifically to an Upper Jurassic population of the species *Holectypus depressus* (Leske), almost all the observations are equally applicable to other species of *Holectypus* and *Caenholectypus*. As few specimens are ever preserved with any spines or pedicellariae attached, only those features that are easily visible on the test are considered.

4.1 Functional morphology of *Holectypus depressus* (Fig. 4.1)

4.1.1 Shape

In profile the test is noticeably flattened, and the height of the test is typically about half of its diameter. The ambitus, which lies relatively low down, is more or less circular in outline. The oral surface is concave and becomes sunken towards the peristome. The low profile to the test could be an adaptation giving stability in currents on either rocky or sedimentary substrata, or could be an adaptation for burrowing. The circular outline might suggest that *Holectypus*, like the regular echinoids, did not have one particular direction of locomotion.

4.1.2 Peristome

The peristome is relatively large, and is positioned centrally on the ventral surface. It is circular in outline and measures approximately one third of the diameter of the test at all sizes (i.e. it grows isometrically). The margin

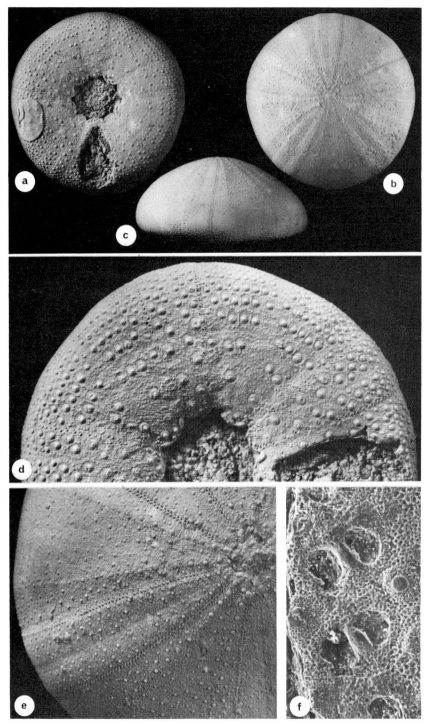

Figure 4.1 *Holectypus depressus* (Leske), an Upper Jurassic holectypoid: (a) oral ($\times$ 1); (b) apical ($\times$ 1); (c) lateral ($\times$ 1); (d) oral surface, anterior to the left ($\times$ 2.5) showing the buccal slits, radial arrangement of tubercles and the lack of ambulacral pore crowding adjacent to the peristome; (e) apical surface ($\times$ 2.5) (f) scanning electron micrograph of two ambulacral pores ($\times$ 110).

of the peristome is notched by prominent buccal slits, proving that *Holectypus* had large compensation sacs that extended out from the peripharyngeal coelom. A well developed perignathic girdle of auricles for the attachment of lantern muscles surrounds the peristome.

The large size of the peristome, the presence of buccal slits and the well developed perignathic girdle all indicate that *Holectypus* had a fully developed lantern throughout life (and indeed lanterns can be found in suitably preserved specimens). Compensation sacs accommodate for pressure changes within the test caused by protrusion and retraction of the lantern, while a large, flexible, peristomial membrane allows the lantern considerable freedom of movement, both laterally and vertically. Where the lantern does not extend out through the peristome, as in clypeasteroids, the peristome is small and lacks buccal slits. Thus *Holectypus* had an active lantern that could be protruded and retracted just as in regular echinoids.

4.1.3 Periproct

The periproct lies on the oral surface, posterior to the peristome. It is oval in outline and unusually large, with a maximum diameter of around 20 to 25% of the test diameter. The periproct is proportionally larger in juveniles (i.e. it grows allometrically), suggesting that its actual size may in some way be crucial. Having the periproct situated orally is an adaptation to prevent fouling the aboral surface with copious faecal discharge. Large volumes of faecal material are produced by deposit feeding and sediment swallowing where the diet has a low organic content. The problem of faecal discharge is particularly crucial to infaunal echinoids, since they cannot rely on surface currents to wash away faecal material. The relatively large size of the periproct (covered in life by a flexible membrane) suggests that particles in voided faecal material may have been fairly large. The allometric growth of the periproct supports this, since juveniles require a proportionally larger periproct than adults feeding on the same sediment. So *Holectypus* was apparently a fairly unselective deposit feeder which, like living cassiduloids, swallowed large quantities of sediment.

4.1.4 Tubercle arrangement

There is an obvious difference between oral and aboral tubercles, both in size and shape, suggesting that the spines on these two surfaces served different functions. Aboral tubercles are small (up to 0.4 mm in diameter) and increase in size only slightly during growth. They are uniformly spaced so that as plates increase in size new tubercles are added at the appropriate distance. Aboral tubercles (and hence spine canopy) density was thus maintained at about one per mm^2 throughout growth. Furthermore, all aboral tubercles are about the same size and new tubercles must grow rapidly at first, and then more or less cease to grow. These observations suggest that it was crucial for *Holectypus* to have an aboral spine canopy of uniform height and density. Such adaptations are found in extant infaunal echinoids where a dense, uniform spine canopy maintains a water-filled space between the test and the spine tips by preventing sediment grains from falling between the spines. A spine canopy density of one per mm^2 is too sparse to have allowed *Holectypus* to burrow in sediments other than coarse sands and shell gravels (see p. 47).

Oral tubercles are very much larger than aboral tubercles, growing up to 1.3 mm in diameter. They are less regularly arranged, often being clumped together, and continue to increase in size during plate growth. All have bilaterally symmetrical areoles for spine muscle attachment. The areole (and hence spine muscle) is greatly enlarged on the adambital side of each tubercle, showing that the power stroke of oral spines must have been directed radially away from the peristome and towards the margin. There-fore *Holectypus*, unlike almost all living irregular echinoids, shows no signs of having had unidirectional locomotion. Burrowing was presumably accomplished by excavating sediment radially out from beneath itself using oral spines, and so would not have involved any forward locomotion.

No concentration of small tubercles is developed around the peristome, so there were no specialised oral spines to help manipulate sediment into the mouth as there are in cassiduloids.

4.1.5   Ambulacral pores

Not only are all five ambulacra identical in *Holectypus* but there is also no differentiation whatsoever in the structure of ambulacral pores from apex to peristome. Ambulacral pores are arranged uniserially in each column, right up to the peristome. Each has two pores separated by a narrow partition and surrounded by a narrow oval platform for the attachment of tube-foot muscle. Similar ambulacral pores in living echinoids are always associated with cylindrical tube feet terminating in a suckered disc. Pre-sumably all the tube feet of *Holectypus* were also simple suckered tube feet.

The absence of any crowding of ambulacral pores adjacent to the peris-tome, and the lack of specialised food-gathering tube feet suggest that, unlike living irregular echinoids, tube feet played almost no part in collect-ing food. As *Holectypus* also lacks oral spines for manipulating food, food must have been gathered by the lantern. Finally, since none of the oral tube feet were particularly muscular they would not have been able to grip in strong current, confirming that *Holectypus* was not a shallow, rocky bottom dweller.

Holectypus had comparatively few tube feet for its size and none were specialised for gaseous exchange: hence its oxygen consumption and metabolic rate must have been relatively low compared with contemporary irregular echinoids such as *Clypeus* and *Nucleolites*, which did have respiratory tube feet. Such a difference could have one of three explanations: (a) *Holectypus* might have inhabited colder waters where metabolic rates are slower; (b) *Holectypus* might have lived epifaunally where the supply of oxygenated water poses no problems; (c) *Holectypus* might have led a comparatively sluggish infaunal life compared to *Clypeus* and *Nucleolites*. The third of these seems most probable since we know *Holectypus* is preserved together with *Clypeus* and *Nucleolites* in sediments that were deposited in warm, shallow-water, reef-associated habitats, and was clearly adapted for burrowing. *Holectypus* then may have led a genuinely sluggish infaunal life. The holectypoid *Echinoneus* provides the best present-day analogy. Like *Holectypus*, *Echinoneus* lacks specialised respiratory tube feet. During the day it stays buried but, unlike cassiduloids which are constantly ploughing through the sediment for food, it remains more or less stationary within the sediment. At night *Echinoneus* comes to the surface to move about and feed when the temperature is lower.

4.1.6   Apical system

Gonopores first appear in the genital plates at about 10 mm test diameter and mark the onset of gonadal development. When they first appear gonopores are very small (0.05 mm in diameter), but they quickly enlarge to about 0.12–0.14 mm by the time the test is about 20 mm in diameter. They remain at this size throughout later growth, indicating that sexual maturity has been achieved. *Holectypus* was thus fully mature and probably spawning by the time it had achieved a test diameter of 20 mm. There is no sexual dimorphism in *Holectypus depressus*, and so the fertilised eggs presumably passed through the normal echinopluteus larval stage.

4.1.7   Plate structure

Interambulacral plates of *Holectypus depressus* are broad but thin, suggesting that they grew rapidly with little provision for plate strengthening. This can be confirmed in suitably weathered or etched specimens that display growth lines. Large individuals show two growth lines on most plates. The first growth line appears in individuals at about 20 mm test diameter, coinciding with the onset of sexual maturity. The second growth line appears at 30–40 mm test diameter and few specimens attained a larger size.

4.2   The palaeobiology of *Holectypus depressus*

Summarising the data presented above, we find that *H. depressus* was an opportunistic species that grew rapidly, attaining sexual maturity towards the end of its first year (15–20 mm diameter) and surviving for a further year or so. It was adapted for life on sedimentary bottoms and burrowed vertically down into coarse sands or shell gravels, using its oral spines to excavate material from beneath itself. It fed using only its lantern to gather food, scooping up bottom sediment fairly unselectively and producing a fairly copious discharge of faecal material. *Holectypus* lacked unidirectional locomotion and, compared with contemporary cassiduloids, had rather a low metabolic rate. It clearly was not adapted for ploughing continuously through the sediment collecting food, as were cassiduloids. It is more probable that *Holectypus* buried itself for protection from predators during daylight hours, remaining relatively inert, and returned to the surface at night to forage.

To this we can add data about the habitat of *Holectypus* from the sedimentary environments in which it is preserved. In England, *Holectypus depressus* inhabited warm, shallow-water, carbonate platform habitats and is most commonly found in protected areas of coarse sediment.

A fairly detailed picture of *Holectypus depressus* has now been built up by considering its functional morphology, and it should be possible to apply a similar procedure to other fossil species. This sort of detailed analysis is most important if we are to try and interpret the biological significance of evolutionary changes that have taken place in fossil lineages.

5 Fossil lineages

The fossil record, despite its drawbacks, remains indispensible in offering the only direct evidence we have that species have changed progressively through time. The actual details of how this change is brought about have, however, been hotly debated in the last decade. The traditional view holds that the great majority of new species evolved slowly through gradual transformation of the entire population (gradualism). The alternative view, that speciation takes place through rapid morphological change in small, isolated populations, the species thereafter remaining relatively immutable until extinction, was termed punctuated equilibrium by Eldredge and Gould (1972). Well documented fossil lineages can not only provide information about the mode and tempo of evolution, but may also give direct evidence for the selection pressures (whether acting at the individual, population or species level) that brought about such changes.

Despite the fact that one of the most widely cited examples of a lineage showing continuous and directional evolutionary changes through time comes from the Upper Cretaceous spatangoid *Micraster*, echinoids provide little support for the gradualistic model of speciation. Even before Eldredge and Gould popularised the idea of punctuated equilibrium, Chesher (1970) had observed that the fossil record of the Tertiary spatangoid *Meoma* followed a pattern of rapid change in an isolated character (forming a new morphospecies), followed by a long period of stasis. Indeed, truly intergrading lineages are few and far between, even in the Chalk, and new species generally appear suddenly in a succession by immigration (as, for example, in the *Infulaster–Hagenowia* lineage outlined below). Even the idea that *Micraster* species gradually evolved from one another has been proved wrong by careful stratigraphical collecting. Stokes (1976) showed that *Micraster decipiens* (*M. cortestudinarium* of old) and *M. coranguinum* co-existed for a period, and so the appearance of *M. coranguinum* is more likely to represent an immigration event following allopatric speciation and cannot be the result of gradual transformation *en masse*. This is not to say that some gradual change does not occur – Ernst (1972) provided several examples. However, these produce only relatively minor changes, largely related to size.

Most of the evolutionary lineages of fossil echinoids with good stratigraphical control prove to consist of a succession of morphologically discrete species rather than a continuous gradation. This in no way detracts from the value of trying to identify the causes for such change, since the processes of natural selection, whether acting on individuals within a population or on populations within a heterogeneous species-cline, produce the same result – namely better adaptation to environmental conditions.

In this chapter I shall outline the changes that took place in four well documented lineages and discuss their possible significance.

5.1 *Linthia–Paraster–Schizaster* lineage

McNamara and Philip (1980) were able to trace the ancestry of the Miocene to Recent spatangoid *Schizaster* back through *Paraster* to the genus *Linthia* on the basis of the progressive morphological changes that they found in the Tertiary schizasterid spatangoids of Australia. Since the Australian fauna of schizasterids consists of stratigraphically isolated finds interpreted as periodical immigration events from more equatorial populations, direct phylogenetic lineage at the species level cannot be proved. However, the morphological trend defined by the succession of incoming species does appear to be valid and can be analysed.

The morphological changes that took place during the evolution of *Schizaster* from *Linthia* are summarised in Table 5.1. McNamara and Philip went on to discuss the functional and palaeobiological significance of these changes, concluding that they represent adaptive changes for deeper burial within finer-grained sediments (Fig. 5.1). The changes in test shape and the increase in plastron spines are clearly adaptations for burrowing in more compact or cohesive sediment. The profile of the test, which in *Linthia* is gently rounded, becomes wedge-shaped in *Schizaster* (Fig. 5.1) to cope with tunnelling through more cohesive sediment. Plastron spines provide the forward thrust in burrowing, and so the progressive increase in plastron spines also suggests that burrowing was becoming more difficult.

The anterior movement of the peristome, the development of a labral projection partially covering the peristome, and the enlargement and deepening of the anterior groove, mark the increasing importance of the anterior groove in feeding. The mouth shifts anteriorly to lie at the base of the groove, while the labral projection effectively blocks the mouth from the sediment directly underneath and makes it face forward. The Recent

Table 5.1 Morphological changes in the lineage *Linthia–Paraster–Schizaster* (from McNamara & Philips 1980).

1 Test becomes taller and longer; dorsal surface becomes anteriorly sloping; tallest part of the test moves posteriorly; apical system moves posteriorly; keel forms in dorsal part of interambulacrum 5.

2 Broadest part of the test moves posteriorly.

3 Anterior groove lengthens and deepens; funnel-building tube feet in ambulacrum III increase in number.

4 Peristome moves anteriorly and labrum expands to project over the peristome.

5 Test increases in size.

6 Plastron becomes broader and more densely tuberculate.

7 Anterior petals become less divergent and more elongate.

8 Gonopores decrease from four to two in number.

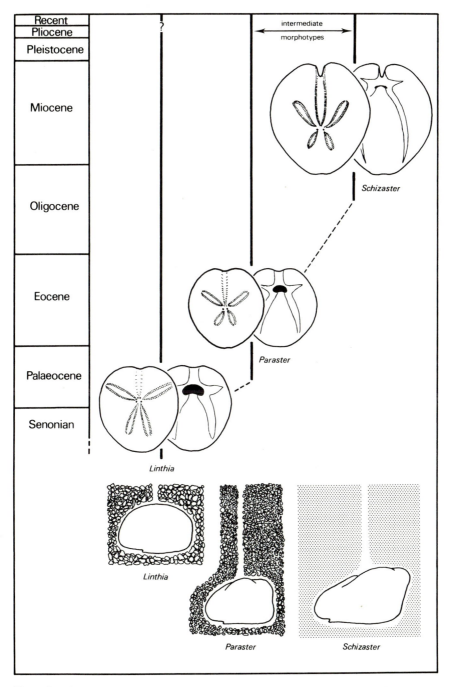

Figure 5.1 Ranges and suggested relationships of *Linthia*, *Paraster* and *Schizaster* in Australia. Profiles show the depth and nature of the sediment inhabited (taken from Mc-Namara & Philip 1980, courtesy of the Australasian Palaeontological Association).

mud-dwelling *Schizaster doederleini* feeds primarily on sediment that falls down into its frontal groove. This sediment is compacted into a loose sediment string aborally and transported down the groove towards the mouth. This feeding mechanism is only practicable, apparently, in medium- or fine-grained sediments, and so these evolutionary changes in the *Linthia–Schizaster* lineage can again be explained as adaptations for life in finer-grained sediments.

The increase in number and density of funnel-building tube feet in the adapical part of the frontal ambulacrum may also be an adaptation for burrowing in finer grades of sediment. However, populations of the spatangoid *Echinocardium* from sands have more funnel-building tube feet than those from muds (Higgins 1975). The greater number of tube feet may therefore be connected with deeper burrowing or the change in feeding technique. Elongation of the anterior petals occurred as a direct consequence of the change in overall shape. The alteration in shape is also presumably responsible for the loss of two of the four gonads by reducing the amount of internal space available. The general increase in size, presumably the result of increased longevity, is common in all lineages where evolution leads from opportunistic generalists to highly adapted specialists.

The changes that occurred during the evolution of *Schizaster* from *Linthia* are all, directly or indirectly, adaptations for living within finer-grained sediments. *Linthia* lived infaunally in the coarser-grained sediments in which it is usually preserved. *Paraster* and *Schizaster* show progressive adaptations for burrowing and feeding in finer, more cohesive sediment, *Paraster* probably living in sands and *Schizaster* in muds. The fact that all three genera have co-existed throughout much of the Tertiary rather than succeeding one another stratigraphically is fairly strong evidence that new morphotypes evolved to fill previously unexploited niches. This is in direct contrast to the next example.

5.2 *Micraster* re-examined

The familiar story of how the spatangoid *Micraster* changed progressively through time within the Chalk facies of England was first outlined by Rowe (1899) and later embellished by Kermack (1954) (Fig. 5.2). It rapidly became one of the classic examples of an evolutionary lineage when Nichols (1959a,b) interpreted these changes as adaptations for deeper burial, using comparisons with extant species. Rowe's phylogeny of *Micraster* remained basically unchallenged until Ernst (1972) and Stokes (1975) published their detailed investigations of *Micraster* based on new material collected with great stratigraphical accuracy (Fig. 5.2). This has shown that the evolution of *Micraster* was in fact much more complex than portrayed by Rowe, and that some species were wrongly identified (Kermack's *M. (Isomicraster) senonensis* is now recognised to be *M. (Gibbaster) gibbus*

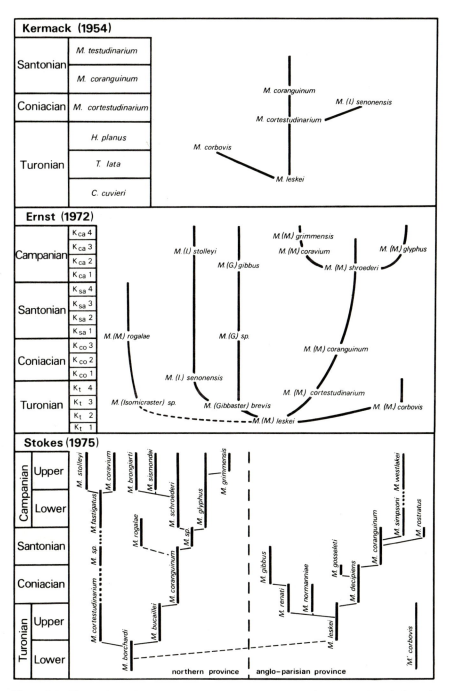

Figure 5.2 Three alternative interpretations of the phylogeny of *Micraster* in Britain and northwestern Europe (adapted from Kermack 1954, Ernst 1972, and Stokes 1975).

Table 5.2 Morphological changes in the lineage *Micraster leskei – M. decipiens – M. coranguinum* (from Kermack 1954 and Stokes 1975).

1	Test becomes taller; dorsal surface becomes anteriorly sloping; tallest part of the test moves posteriorly; apical system moves posteriorly; dorsal part of interambulacrum 5 develops into a keel.
2	Test becomes broader, the broadest part of the test moves posteriorly.
3	The anterior groove lengthens and deepens and becomes more densely tuberculate.
4	The peristome moves anteriorly; labrum enlarges and projects over the peristome; labral tubercles become denser.
5	Test increases in size and thickness.
6	Madreporite increases in size at the expense of other genital plates.
7	Petals increase in length and develop granular ornamentation.
8	Periplastronal ornamentation becomes progressively stronger.
9	Subanal fasciole becomes broader.

and Rowe's *M. cortestudinarium* is *M. decipiens*). Details of the taxonomy and phylogeny within the genus *Micraster* are still unsettled. However, all agree that there was a direct line of descent leading from *M. leskei* through *M. decipiens* to *M. coranguinum*. This is unlikely to represent a gradualistic evolution since *M. decipiens* and *M. coranguinum* overlap in range and species succeed one another with little integration.

Table 5.2 lists the morphological changes that occurred in the *M. leskei–M. coranguinum* lineage. The similarity between the evolution of *Micraster* and evolution in the *Linthia–Schizaster* lineage is very striking (compare Tables 5.1 & 2). Points 1–5 are identical in both lineages and carry the same functional implications. The change in shape was probably an adaptation for burrowing in more cohesive sediment. The lengthening and deepening of the frontal groove, its increased tuberculation, the anterior shift of the peristome and the development of a projecting labrum are all attributable to a change in feeding technique, with greater emphasis being given to material transported down the frontal groove. The increasing size of the madreporite may correlate with increasingly active funnel-building tube feet but data on living spatangoids are lacking. Petals enlarge to house more respiratory tube feet and the ornamentation within the petals pre-sumably acted to increase the area of ciliated epithelium and so improve ciliary currents past the tube feet. Oxygen uptake was obviously critical to *Micraster*. Finally, the subanal fasciole, which in extant spatangoids helps to draw water through the burrow, progressively broadens in *Micraster*. This may simply reflect the increasing size of *Micraster* and the proportion-ally greater volume of water to be drawn through the burrow.

The morphological changes within the *Micraster* lineage are all adapta-tions for burrowing and feeding in fine-grained sediments. Nichols (1959b) suggested that, during the evolution of *Micraster*, a change in niche took place within an effectively constant environment and that species burrowed progressively deeper through time. However, this does not explain why evolution in *Micraster* should compare so closely with evolution in the *Linthia–Schizaster* lineage, which was brought about by a change of habitat from coarse sands to muds. Furthermore, species of *Micraster* did not occupy different niches but replaced one another in stratigraphic sequence and were probably in direct competition. The significance of these observations becomes apparent when the ancestry of *Micraster* and the palaeoenvironmental setting at the start of the Upper Cretaceous are considered.

Micraster leskei appears to have evolved from the genus *Epiaster* via *E. michelini* (Stokes 1977). *Epiaster* is a common spatangoid that first appeared in the Albian. In shape it resembles *Linthia* and, like *Linthia*, it is found predominantly in sand lithofacies. *Epiaster* was morphologically well enough adapted to have lived infaunally within sands and to have produced respiratory and subanal tunnels. In northwestern Europe, chalk facies started to develop during the Cenomanian, marking an important change in palaeoenvironment. As the seas continued to deepen, previously extensive sand facies were replaced by fine-grained chalk marl, producing a faunal crisis. Up to this time the great majority of infaunal spatangoids appear to have been adapted for life in coarse sediments. The success of *Micraster* in the Upper Cretaceous was due to its ability to exploit the newly developed chalk marl facies.

Although *Epiaster* was probably never able to burrow within the fine-grained chalk sediments, *E. michelini* did migrate into areas of chalk deposition at the start of the Turonian, where it gave rise to the infaunal *Micraster leskei*. This breakthrough into a new habitat led to rapid diversification of *Micraster* as it expanded in range. However, this in turn must have led to competition for resources, and we find a progressive improvement through time, not necessarily for deeper burrowing, but just for more efficient burrowing and feeding within this habitat. Thus early morphotypes that were relatively inefficient were replaced by better adapted morphotypes through interspecific competition (Fig. 5.3).

This model explains some rather peculiar features of *Micraster* that were rather puzzling. *Micraster* solved the problems of living infaunally in fine sediments rather differently from other spatangoids. To prevent fine sediment particles from falling through the spine canopy and suffocating the animal, extant mud-dwelling spatangoids have dense aboral spines and an aboral fasciole. The aboral fasciole produces a mucous envelope that covers the spine canopy and that stops fine particles falling between spines. Surprisingly, *Micraster* has no aboral fasciole and, although aboral spines were spatulate tipped, they were unusually sparse (less than 1 per mm^2 as opposed to 4–16 per mm^2 in other mud-dwelling spatangoids). *Micraster* differs from other spatangoids, however, in having remarkably dense aboral miliary tubercles and it seems likely that it was these miliary spines that formed the protective canopy against fine sediment. As some advanced species of *Micraster* have the beginnings of a peripetalous fasciole, developed from aboral miliaries, all miliary spines may have been

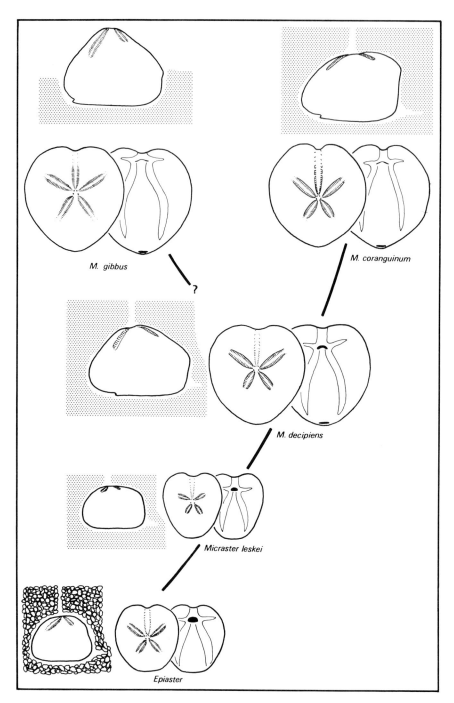

Figure 5.3 Inferred phylogeny of the *Micraster* lineage in southern England. Profiles show the depth of burial and the nature of the sediment inhabited.

able to secrete mucus for a protective coat, even in early species. *Micraster* inherited its aboral spatulate spines from ancestors that lived infaunally within coarse sediments where the density of its spiny canopy would have been effective.

The success of *Micraster* in invading and dominating the chalk seas is owed largely to the evolutionary innovation of a dense miliary spine canopy that allowed it to live infaunally. Following the initial invasion of the new habitat, competition and selection between populations led to the progressive modification of a 'sand-dwelling' morphotype into a 'mud-dwelling' morphotype.

5.3   The *Infulaster–Hagenowia* lineage

The Upper Cretaceous holasteroid *Hagenowia* is one of the most immediately recognisable of Chalk echinoids from northwestern Europe, because of the way in which the apical part of the test is drawn out into a slender rostrum. The evolution from the fairly 'normal' *Infulaster excentricus* to the extremely bizarre *Hagenowia blackmorei* is well documented and involved a number of intermediate species (Fig. 5.4). These species are well delineated and show little integration in southern England. Successive species in the main lineage do not overlap in stratigraphical range. The profound changes that occurred in the evolution of *Hagenowia* have been described and interpreted by Gale and Smith (1982) and the more important ones are listed in Table 5.3.

Most of the evolutionary changes that occurred are associated with the development of the rostrum and its increasingly important role in feeding. For example, elongation of the rostrum led to an overall reduction in

Table 5.3 Morphological changes in the lineage *Infulaster excentricus – Hagenowia blackmorei* (from Gale & Smith 1982).

---

1 Size decreases and test becomes shorter and taller.
2 Rostrum develops through apical elongation producing a split apical system and loss of ambulacral columns.
3 Rostrum becomes more vertical and more slender as well as developing a stronger cross-sectional shape.
4 Anterior sulcus becomes more shallow; sulcus becomes enclosed by a dense arch of spines.
5 Aboral respiratory tube feet are lost.
6 Peristome moves from ventral to anterior face and plastron becomes interrupted; phyllode pores decrease in number and are lost; peristome becomes fully enclosed within the tunnel formed by the anterior sulcus and its arch of spines.
7 Gonopores reduced from four to two.
8 Dorsal spines become better adapted for burrow maintenance; anterior scraping spines become better developed.
9 Change from a double to a single subanal tuft of spines.
10 Tubercles and ambulacral pores of the anterior sulcus become differentiated near the apex.

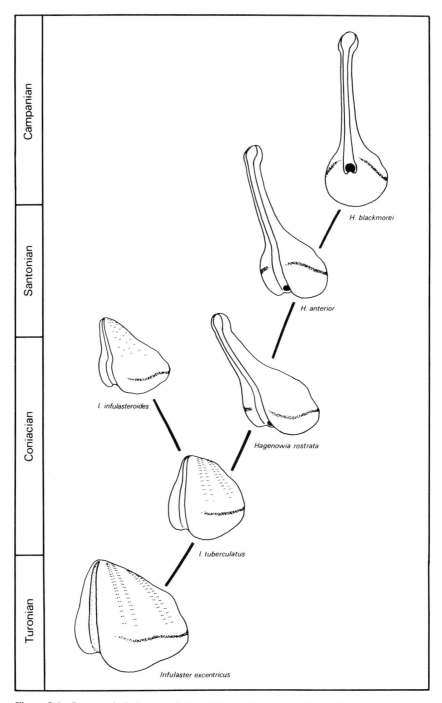

Figure 5.4 Suggested phylogeny of the *Infulaster–Hagenowia* lineage (not to scale).

internal space for gonadal growth and two of the four gonads were lost. Ambulacrum III runs the entire length of the rostrum and is sunken, forming an anterior sulcus. That this sulcus was an important passageway for food transport is quite clear from the various adaptive changes that took place. The mouth, which is ventral in *Infulaster*, shifted to the anterior edge of the ventral surface in *Hagenowia rostrata* and is situated well up the anterior face in *H. blackmorei*. In this position it lies at the base of the anterior sulcus well away from the floor of the burrow. Accompanying this change there was a reduction in number of phyllodal ambulacral pores surrounding the peristome, leading to their complete absence in *H. blackmorei*. This suggests that *Hagenowia* abandoned using penicillate tube feet to collect sediment and came to rely solely on material coming down the anterior sulcus. Furthermore, a dense archway of spines developed across the anterior sulcus effectively making a tunnel that led directly to the peristome and opened only at the apex of the rostrum. Within the anterior sulcus, miliary tubercles and spines became particularly dense near the apex, presumably to create stronger ciliary currents, and the most adapical tube feet became differentiated, apparently to fulfil a sensory role.

The rostrum, which was initially oblique, became more vertical as well as more robust in cross section despite its increasing slenderness. This, in effect, placed the main body of the test deeper within the sediment, and so the spines became better adapted for excavating and maintaining a burrow. Unlike the two previous examples, there was a dramatic decrease in size during the evolution of *Hagenowia*. The precise reason for this is uncertain, but it resulted in the subanal spines coalescing from a double to a single tuft. It may also help to explain why aboral respiratory tube feet were lost in *Hagenowia*, since in smaller animals a greater proportion of oxygen can be obtained by direct diffusion through the body wall.

Gale and Smith (1982) built up detailed arguments about the functional significance of these evolutionary changes and concluded that neither *Infulaster* nor *Hagenowia* could have lived fully infaunally within chalk marl, since they lacked funnel-building tube feet. Both apparently ploughed through the sediment with their apex at the surface. The evolutionary development of *Hagenowia* is seen as adaptation to a new feeding technique. It is suggested that *Infulaster excentricus* fed principally upon the sediment beneath itself, supplemented by a certain amount of material falling down the anterior sulcus. *Hagenowia* may have evolved to feed solely upon the surface detritus layer that was cascaded down the anterior sulcus as the echinoid ploughed its way through the sediment. As the surface layer of sediment usually has the highest organic content, *Hagenowia* was apparently adapting to less nutrient-rich sediments by becoming more efficient and selective in its feeding. However, rostral elongation also made *Hagenowia* less obvious from the surface which may have been important for avoiding potential predators.

A possible reason why *Hagenowia* should have evolved its highly specialised feeding technique can be found by considering the prevailing palaeogeographical setting. During the Upper Cretaceous there was a progressive rise in sea level, flooding the continental shelf seas to depths of up to 800 m (Hancock and Kauffman 1979). As water depth increases, the amount of nutrients reaching the sea floor generally decreases, and so extant deep-sea echinoids such as *Pourtalesia* feed by skimming off the surface layer of detritus-rich sediment. It is possible that *Hagenowia* adopted a similar feeding strategy in response to reduced levels of nutrient input brought about by ever-deepening conditions.

5.4   The evolution of primitive irregular echinoids

For the final example of an evolutionary lineage I want to consider the changes that took place during the initial evolution of irregular echinoids. This is somewhat more speculative than previous examples, since no firm line of descent can yet be proved. However, it is possible to construct a sequence of Lower Jurassic species that shows a morphological progression leading from the regular echinoid *Eodiadema* to the cassiduloid *Clypeus* via '*Plesiechinus*' *hawkinsi*, *Eogaleropygus* and *Galeropygus*. Kier (1982) has outlined the steps that must have occurred to transform a regular into an irregular echinoid and has made functional interpretations. The order in which these changes took place is highly significant as it records how echinoids achieved the breakthrough into a major new niche.

Eodiadema is chosen as the regular echinoid from which later irregular echinoids might be derived because it possesses one important morphological feature that identifies it as a stem member of the Irregularia, namely teeth diamond-shaped in cross section. This type of tooth probably evolved in *Eodiadema* or some closely related genus through paedomorphism. In diamond-shaped teeth, side processes are formed precociously before primary tooth plates have begun to elongate and give the tooth a self-sharpening structure at a much smaller size than is possible in other types of tooth. This was vital for an echinoid of such small size as *Eodiadema* which, fully grown, did not reach more than 1 cm in diameter. Diamond-shaped teeth then presumably had no functional advantage over grooved teeth but evolved as a consequence of extreme paedomorphic development in *Eodiadema*. Irregular echinoids later inherited and modified the design of this tooth. Functional analysis suggests that *Eodiadema* was an epifaunal opportunist living in temporarily hospitable sedimentary environments and was probably fairly omnivorous in its diet.

Several important changes had taken place by the time the Sinemurian '*Plesiechinus*' *hawkinsi* appeared. The test had become flattened in profile (Fig. 5.5) which was probably an adaptation for living on loose, unconsolidated sediment. The low profile would have provided greater stability in currents (since tube feet could not be used for anchorage) and brought a larger proportion of spines in contact with the sea floor for more efficient

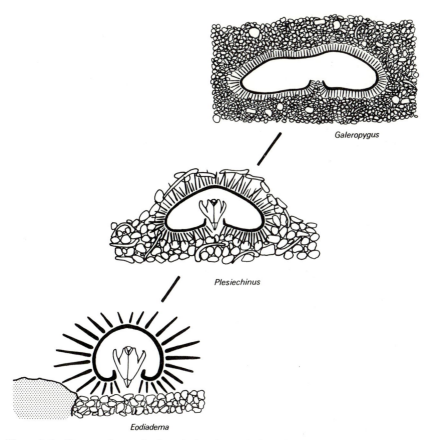

Galeropygus

Plesiechinus

Eodiadema

Figure 5.5 Changes that took place during the evolution of irregular echinoids from their regular ancestor (not to scale) (adapted in part from Smith 1978a).

locomotion. There were also changes in the size and arrangement of tubercles and spines. Instead of having just one large tubercle and spine per plate there were now some three to five small tubercles and spines. Aboral tubercles were not yet particularly dense nor uniform in size and so there was no uniform canopy of spines. All the tube feet were apparently suckered, like those of *Eodiadema*. *Plesiechinus* then does not seem to have been particularly adapted for an infaunal mode of life and probably lived epifaunally or semi-infaunally (Fig. 5.5). Extant regular echinoids with short aboral spines and suckered tube feet generally cover themselves with loose material and *Plesiechinus* probably did the same. '*P.*' *hawkinsi* almost certainly still had a fully functioning lantern, and its periproct lay fully enclosed within a complete apical system. It is therefore unlikely to have fed by swallowing sediment wholesale and presumably still had a diet of relatively high organic content.

With the evolution of *Eogaleropygus*, further important changes had taken place. *Eogaleropygus* appears to have lost its lantern, at least as an adult, and the peristome had consequently become smaller. Ambulacral pore morphology suggests that the tube feet had started to differentiate, those aborally becoming thin-walled for gaseous exchange whereas the oral tube feet remained suckered and had become slightly more concentrated adjacent to the peristome. Tubercles (and presumably spines) had become smaller and more densely packed, particularly aborally where they were all more or less the same size. *Eogaleropygus* quite clearly fed on bottom material that it collected and transferred to its mouth using oral tube feet. It was also functionally well enough adapted to have lived fully infaunally within relatively coarse sediments. However, the anus still opened at the apex of the test and the copious amount of faeces produced from deposit feeding must have posed something of a problem, and probably led to the evolution of *Galeropygus*.

In *Galeropygus* the periproct is still situated close to the apex but it opens into a long anal sulcus that channelled faeces posteriorly away from the aboral respiratory surface and into the animal's wake. *Galeropygus* was well adapted for burrowing (Fig. 5.5) and there is a bilateral symmetry to oral tubercles that suggests that *Galeropygus* ploughed through the sediment with ambulacrum III foremost.

The final changes that took place leading to the evolution of *Clypeus* from *Galeropygus* involved improvement of oral phyllodes (increasing the number of food-gathering tube feet), the appearance of bourrelets (increasing the number of manipulative spines adjacent to the mouth) and the development of petaloid ambulacra and tube feet highly adapted for gaseous exchange (presumably allowing *Clypeus* to become a more active burrower).

If this evolutionary scheme is more or less correct then cassiduloids evolved from regular echinoids through the following series of steps:

(a) development of diamond-shaped teeth through paedomorphism;
(b) changing to living in areas of unconsolidated sediment;
(c) covering and later burrowing into sediment for protection or stability (concurrent with (d));
(d) changing to a diet of sediment, collected at first using the lantern but later using only tube feet;
(e) burrowing continuously through the sediment in a more systematic strategy of feeding.

6 Spatial distribution and palaeoecology

Echinoids are now found in almost all marine habitats, from polar to equatorial regions and from the intertidal zone to depths greater than 7000 m. Some species reach almost cosmospolitan distribution but most are geographically restricted, and all are habitat restricted to a greater or lesser degree. In this chapter, first the factors that influence the distribution of present day echinoids will be examined, and then some examples where spatial distribution has changed through time will be discussed. The chapter ends with a selection of examples to show how echinoids can be used as palaeoenvironmental indicators.

6.1 Controls on distribution

Geographical barriers (land barriers, wide oceanic basins, etc.) and the oceanic current pattern act as the primary controls on species distribution in echinoids (Mayr 1954). However, even within a small area, echinoid distribution is notoriously patchy. There are many reasons for this and work by Kier and Grant (1965) and Ebert (1971) has identified eight factors that can affect the distribution pattern of a species, each of which is discussed below.

6.1.1 Substratum

The nature of the substratum (and by this I mean not just grain size but sediment stability, degree of sorting, content of organic material, porosity and permeability) is arguably the single most important factor influencing local distribution. Echinoid larvae are well known to show a strong preference and can delay settlement until a suitable substratum has been found. In almost all species, a strong correlation thus exists between adult distribution and the nature of the sea floor. By and large, regular echinoids live on rocky bottoms or stable sediments, whereas irregular echinoids live on or in unconsolidated sediments, though there are many exceptions to this.

Irregular echinoids are usually much more restricted in their choice of substratum than regular echinoids. This is because irregular echinoids are morphologically often highly adapted to cope with burrowing and feeding within very specific grades of sediment. For example, McNulty *et al.* (1962) found that around Florida, USA the spatangoid *Moira atropos* was only found in sediments consisting of 75–90% sand-grade particles with 0–12%

gravel and 10–20% silt, and mellitid sand-dollars live only in medium- to fine-grained sands. Reviews of substratum specificity in irregular echinoids have been given by Lawrence and Ferber (1971) and Smith (1980a).

Amongst the regular echinoids, Ernst (1973a) found that the distribution of the Mediterranean species of *Arbacia*, *Paracentrotus* and *Sphaerechinus* was strongly correlated with the type of substratum, and Heatfield (1965) was able to prove that adult regular echinoids are clearly influenced by the texture of the bottom. Even in areas of otherwise uniform rocky bottom, Shepherd (1973) has suggested that the distribution of certain regular echinoids is restricted by the availability of suitable crevices in which to live.

Fossil species usually show a similar restriction to specific lithofacies. Different lithofacies often contain quite different echinoid faunas which must be, at least in part, due to substratum differences. For example, the Santonian (Upper Cretaceous) of northern Germany includes both calcarenite and chalk–marl lithofacies of equivalent age but with quite distinct echinoid faunas (Ernst 1973b).

6.1.2   Hydrodynamic regime

Current velocities and the extent of water turbulence directly influence the distribution of echinoids. Several groups of regular echinoids are adapted for life in the high energy surf zone. They can only survive in this environment by keeping a firm grip of the substratum and are thus restricted to rocky bottoms and must possess strong suckered tube feet. Few irregular echinoids can live in unstable sediments within the surf zone except for deep-burrowing spatangoids and sand-dollars. Indeed, mellitid sand-dollars are largely restricted to sediments within the surf zone (possibly because there is no competition from other echinoderms there) although they too are excluded from areas with too much turbulence (Weihe & Gray 1968).

The planktonic larvae of echinoids are transported in oceanic currents and so the hydrodynamic regime also directly controls larval distribution. A rather perplexing feature of the present-day echinoid biogeography is that the fauna of the Gulf of Guinea (West Africa) is surprisingly similar to the fauna of the West Indies. This cannot be solely due to continental drift as the fauna is relatively young compared to the age of the Atlantic. Migration of species appears to have been from west to east across the Atlantic, but cannot have occurred by 'island hopping' since the faunas on either side of the Atlantic are more similar to one another than to the fauna of mid-Atlantic islands (Chesher 1966). Nor does it seem possible that pelagic larvae could survive the year or so required to cross the Atlantic via the northern Gulf Stream. The discovery of a relatively rapid equatorial undercurrent running from Brazil to West Africa and taking just two to three months to cross seems to provide the explanation for the faunal similarity.

6.1.3   Predation

Predation usually has little effect on the structure of an established population or on its distribution (Hendler 1977). Birkeland and Chia (1971) did, however, suggest that predation by asteroids might prevent the sand-dollar *Dendraster* from occupying physiologically acceptable subtidal habitats.

6.1.4   Salinity

Although echinoids are generally considered to be stenohaline, some species are able to acclimatise to hypersaline or hyposaline conditions. Populations of *Mellita* and *Psammechinus* are known from hyposaline habitats with salinities as low as $20^{\circ}/_{\circ\circ}$ (Gezelius 1963, Durham 1966) and, in the Persian Gulf, echinoids were found at salinities of up to $45–8^{\circ}/_{\circ\circ}$ by Clarke and Keij (1973). Species diversity, as might be expected, drops off rapidly with any departure from normal salinity.

6.1.5   Temperature

Temperature can affect echinoid distribution in two ways. First, it can control the upper limit of shallow-water populations since exposure during daylight hours may result in mass mortality (Khamala 1971). Shore populations are therefore restricted to permanent deep pools or to the low intertidal zone. Secondly, cold oceanic currents can act as barriers to dispersal and have been invoked as limiting the northern distribution of the sand-dollar *Mellita* and other Caribbean echinoids.

6.1.6   Food availability

Regular echinoids are so flexible in their diet that they are rarely restricted in their distribution by the availability of one particular food. However, a fair proportion of regular echinoids do appear to be restricted principally to the shallow-water photic zone where plant and algal primary production occurs. The restriction of crevice and burrow-dwelling species to the uppermost few metres of water may be because they rely on capturing drifting algae (Campbell *et al.* 1973). Food availability is difficult to assess for irregular echinoids but the common clumped distribution of populations may reflect local differences in the organic content of the sediment.

6.1.7   Depth

Depth *per se* has little direct control over species distribution (Ebert 1971) although many species appear to be restricted by depth. This is because depth influences many factors such as the amount of turbulence, type of substratum, light (and hence the quantity of plant and algal growth) and temperature, all of which may themselves directly influence echinoid distribution. There are some species of regular echinoid that live only within the top few metres of water, but this is probably because they are highly adapted for life in these very turbulent habitats where they can avoid competition with other species. There are also several echinoid groups that are known only from deep-water habitats, notably echinothurioids, holasteroids, pourtalesiids and the spatangoid families Urechinidae, Aeropsidae and Asteromatidae.

6.1.8 Behaviour

The daily migration between areas of food and areas of shelter known to occur in some shallow-water regular echinoids is one example of a behavioural trait that affects local distribution.

6.1.9 Chance

Ebert (1971) found some inexplicable distribution patterns amongst the shallow-water echinoids that he studied – he attributed these to chance.

6.2 Species distribution and biogeographical realms

Echinoid species and most genera are geographically restricted to some extent. As in other marine groups, species diversity for echinoids is highest in tropical and subtropical shelf seas and decreases markedly with increasing latitude and increasing depth. Of the 50 or so families of echinoids found today, only six are represented in the Antarctic and only 44 species (less than 5% of all living species) live in Antarctic waters, the majority belonging to just five genera. A similar decrease in diversity towards polar regions is quite evident in past geological periods.

The decrease in diversity with increasing depth is equally as pronounced, even in relatively shallow waters (*vide* Ebert 1971). Only a small proportion of species of major groups living in continental shelf seas range into the deeper waters of the continental slope. Archibenthal faunas (in depths from 200 m to about 1000 m) are largely made up of cidarids, diademataceans and spatangoids. At greater depths, endemic deep-water groups such as echinothurioids, holasteroids, pourtalesiids and some spatangoid families dominate. Diversity, though, is low and progressively decreases with increasing depth so that the fauna of abyssal plains is almost entirely made up of pourtalesiids, and only five genera of echinoids live at depths greater than 4500 m.

The restriction of species and genera to particular geographical areas has been used to define biogeographical provinces and realms (Hyman 1955). Each province has its own distinctive suite of species endemic to that region, and provinces with similar faunas are grouped into faunal realms. Faunal realms have been just as distinctive in the geological past, although surprisingly little has been written about the distribution of fossil echinoids on a global scale. Probably the most comprehensive palaeobiogeographical study using echinoids that has been published was that on Cretaceous faunas by Stokes (1975). Stokes based his analysis primarily on spatangoid distribution and was able to give a very detailed account of the provinces and subprovinces of western Europe, defined on the species of *Micraster*. Palaeobiogeographical boundaries changed in position through time, particularly across Europe where there was no geographical barrier to migration (Fig. 6.1). Unfortunately, outside Europe, faunas are often less well known and changes cannot be followed in the same sort of detail.

During the Cretaceous there was a more or less unobstructed circum-equatorial seaway allowing genera to become widely distributed. Despite this, endemism at the species level was common, particularly towards the end of the Cretaceous when faunal provinces became most pronounced. Six important centres of endemism can be recognised (Fig. 6.2), four of which can be divided into a number of provinces. These are as follows:

(a) *Gulf and Caribbean centre.* The rich and diverse Cretaceous faunas of Mexico, Cuba and the Gulf states of the USA are well documented. The Lower Cretaceous fauna is very similar to that of the circum-Mediterranean region at the generic level, but lacks the diversity of cassiduloids and holasteroids. The fauna became more distinctive during the Upper Cretaceous, particularly amongst cassiduloids. The Gulf and Caribbean fauna extended well into South America during the Lower Cretaceous, reaching as far as Brazil on the east coast and Peru on the west coast, though with high levels of species endemism and greatly reduced diversity. In the Upper Cretaceous of Brazil, typical Gulf species occur together with North African cassiduloids indicating a mixed Gulf and circum-Mediterranean influence. The Upper Cretaceous marine transgression extended the Gulf fauna northwards into North America but again with much reduced diversity and species endemism. In the Western Interior, Gulf species are progressively lost northwards but extend up as far as Alberta, Canada.

(b) *Circum-Mediterranean centre.* The echinoid fauna of North Africa and southern Europe was both abundant and diverse. This region was probably the most important centre of diversification, particularly during the Lower Cretaceous, and both holasteroids and spatangoids appear to have originated here. The northern boundary of this region is hard to define, since typical north European forms migrated southwards and *vice versa* to produce a mixed faunal zone. It is clear that the boundary could not have been a topographical barrier and was

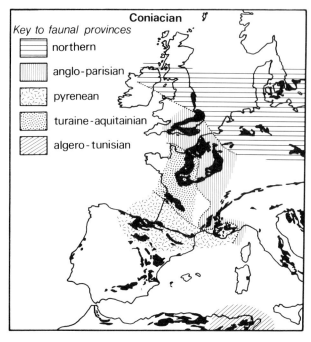

Figure 6.1 Faunal provinces in western Europe and North Africa during the Coniacian and Campanian (Upper Cretaceous) established using *Micraster*. Areas shaded in black indicate present-day outcrop (from Stokes 1975, courtesy of the author).

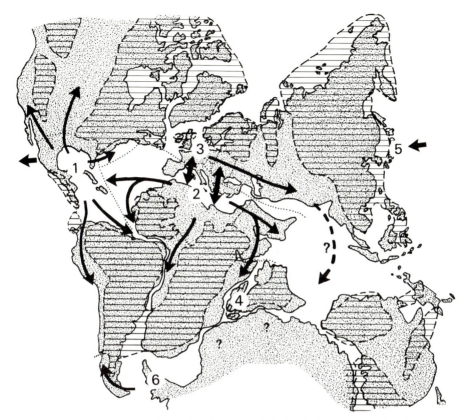

Figure 6.2 Major centres of endemism for echinoids during the Cretaceous and their spheres of influence. (1) Gulf and Caribbean; (2) circum-Mediterranean; (3) North European; (4) Indo-Madagascan; (5) Japanese; (6) Antarctic. Cretaceous land masses hatched. (Palaeogeographical reconstruction taken from Smith *et al*. 1973, courtesy of the Palaeontological Association.)

probably related to temperature and oceanic current regime. Elements of the circum-Mediterranean fauna extended as far north as southern England at times. To the south, the African landmass formed the southern limit, although a typical North African fauna (but with high species endemism) reached Angola in the Albian. With the opening of the South Atlantic in the Upper Cretaceous the West African fauna became more mixed with both Gulf and North African elements. To the east the circum-Mediterranean fauna extended as far as Palestine and Turkey. Further east, in Iran, Saudi Arabia and Pakistan, the fauna, though still showing strong Mediterranean affinities, had a high proportion of endemic species and a number of endemic genera.

(c) *North European centre.* During the Lower Cretaceous, this region simply had a rather depleted circum-Mediterranean fauna though with a certain amount of endemism at the species level. In the Upper

Table 6.1 A selection of the more distinctive echinoid genera for each of the Cretaceous faunal realms. Widespread genera, that are common to three or more regions are not included.

	Gulf and Caribbean	Circum-Mediterranean	North European	Indo-Madagascar	Japanese	Antarctic
Regular echinoids	*Boletechinus* *Dumblea* *Eurysalenia* *Goniophorus* *Goniopygus* *Loriolia* *Orthopsis* *Rachiosoma*	*Codiopsis* *Diplopodia* *Echinotiara* *Gauthieria* *Heterodiadema* *Noetlingaster* *Orthopsis* *Polydiadema* *Porosoma* *Prionocidaris* *Rhabdocidaris* *Sardocidaris* *Temnocidaris* *Tetragramma*	*Araeosoma* *Asthenosoma* *Cottaldia* *Echinothuria* *Glyphocyphus* *Glyptocyphus* *Goniophorus* *Hyposalenia* *Phormosoma* *Tylocidaris* *Zeugopleurus*	*Gauthieria* *Gomphechinus* *Goniopygus* *Micropedina* *Orthopsis* *Tetragramma*	*Pseudocidaris*	*Cyathocidaris*
Cassiduloids	*Hardouinia* *Lefortia* *Parapygus* *Domechinus*	*Archiacia* *Astrolampas* *Claviaster* *Clypeolampas* *Clypeopygus* *Eurypetalum* *Faujasia* *Gentilia* *Oolopygus* *Petallobrissus* *Pygorhynchus* *Pyguropsis* *Pygurus*	*Catopygus* *Plagiochasma* *Ochetes* *Trematopygus*	*Australanthus* *Botriopygus* *Clypeolampas* *Gongrochanus* *Stigmatopygus*	—	—

Table 6.1 *Continued*

	Gulf and Caribbean	Circum-Mediterranean	North European	Indo-Madagascar	Japanese	Antarctic
Holectypoids	—	*Rhynchopygus* *Zumoffenia*	*Echinogalerus* *Galerites* *Globator*	—	—	—
Diasteroids	—	—	*Acrolusia* *Tithonia*	—	—	—
Holasteroids	—	—	*Hagenowia* *Infulaster* *Offaster* *Sternotaxis*	*Basseaster* *Hemipneustes* *Pseudoholaster*	—	*Nordenskjoldaster*
Spatangoids	*Heteraster* *Leiostomaster* *Palhemiaster* *Pliotoxaster* *Proraster* *Washitaster*	*Douvillaster* *Enallaster* *Enallopneustes* *Heteraster* *Miotoxaster* *Palhemiaster* *Pliotoxaster* *Polydesmaster* *Proraster* *Toxaster*	*Gibbaster* *Isomicraster*	*Distefanaster* *Homoeaster* *Mokotibaster* *Toxopatagus*	*Aphelaster* *Paraheteraster* *Cottreaucorys* *Nipponaster* *Pseudowashitaster*	*Vomeraster*

Cretaceous it developed its own distinctive fauna which expanded south-wards over much of Europe. The North European fauna extended from Britain eastwards at least as far as Turkistan and the Caucasus, but western Europe, eastern Europe and asiatic Russia all have their own endemic species. Mixed European and Mediterranean faunas have been recorded from Hungary, northern Turkey and France.

(d) *Indo-Madagascar centre.* Like the North European region, the Indo-Madagascar region did not initially have its own distinctively endemic echinoid fauna. Lower Cretaceous echinoids of northwestern India have a decidedly European affinity, and a similar though reduced fauna is present to the south of Madras. In contrast, the Lower Cretaceous fauna of East Africa and Madagascar has close ties to the North African fauna, and presumably a major barrier must have existed between Madagascar and southeastern India at this time. A distinctive fauna began to develop in southern India and Madagascar towards the end of the Cretaceous, though this still retained a strong North African affinity.

(e) *Japanese centre.* Japan has a distinct but very low-diversity echinoid fauna, possibly from deep water. There are only two known species of regular echinoid in the Cretaceous and the overwhelming majority of irregular echinoids are spatangoids. Although it is difficult to compare this fauna with that of other areas, the spatangoids suggest some affinities with North American faunas.

(f) *Antarctic centre.* In Patagonia and the adjacent areas of Antarctica there is a low-diversity echinoid fauna consisting predominantly of the cidarid *Cyathocidaris* and a few holasteroids. This fauna extended up the east and west coasts of South America, where a certain amount of mixing with Gulf and Caribbean forms took place.

Data on echinoid faunal provinces during the Cretaceous is summarised in Figure 6.2, and Table 6.1 lists a selection of representative genera for each centre. The apparent low endemism above species level at this time is probably due largely to the presence of the circum-equatorial seaway during the Cretaceous that allowed much more faunal mixing. As today, highest diversity Cretaceous faunas were present in tropical and subtropical regions and diversity decreased noticeably towards the poles. Many province boundaries were gradational both spatially and through time, suggesting that they were defined by the pattern of oceanic currents that existed rather than by topographical barriers.

6.3 Evolution in time and space

Recognising biogeographical realms in the geological past is just one of several uses to which detailed knowledge of species distribution can be put.

Foster and Philip (1978), for example, have used the timing of the appearance of Tertiary echinoids in Australia and New Zealand to identify migration pathways and reconstruct palaeogeography. Another approach that can be very informative is to follow the diversification of a single clade through time, as was done for saleniids by Durkin (1980) and sand-dollars by Seilacher (1979). Finally, one can follow the way in which one particular habitat is invaded and filled, and as an example of this I shall look at the problem of the origin of the deep-sea fauna.

6.3.1 Tertiary migration pathways between Australia and New Zealand

As Australia and New Zealand shared a similar echinoid fauna throughout the Tertiary, Foster and Philip (1978) were able to use the distribution of species and genera through time to identify the direction and timing of migration within this area. Sea-floor spreading in the Tasman Sea was more or less complete by 60 million years ago so that, throughout the Tertiary, Australia and New Zealand were as widely separated as they are

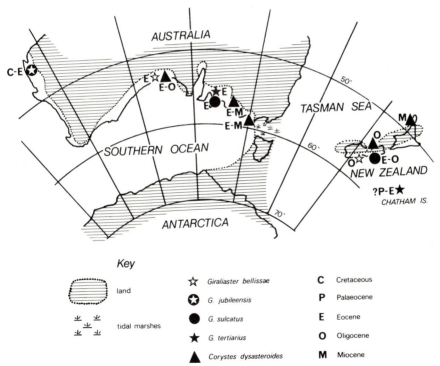

Figure 6.3 Palaeogeographical reconstruction of Australia, New Zealand and part of Antarctica during the Oligocene with known occurrences of holasteroid echinoids and their ages (from Foster & Philip 1978, courtesy of the Palaeontological Association).

today. At the start of the Tertiary, Australia and Antarctica were still connected by a land bridge and it was believed that the Southern Ocean that now separates them did not open fully until the Lower Oligocene. Foster and Philip were able to prove that the migration of echinoids had consistently been from west to east throughout the Tertiary. First, they identify a faunal link between Australia and southeastern Africa by noting that the Australian Tertiary holasteroids are closely related to the Upper Cretaceous genus *Basseaster* from Madagascar, and the typically South African spatangoid *Spatangobrissus* appears in the Miocene of western Australia. Secondly, the holasteroid *Giraliaster* first appeared in the uppermost Cretaceous of western Australia but did not reach southern Australia until the Eocene (Fig. 6.3). Thirdly, genera and species that are common to both Australia and New Zealand consistently appeared earlier in Australia (Table 6.2). Foster and Philip concluded that, since migration into New Zealand was already occurring by the Eocene, the Tasmanian high must have been breached and a marine pathway between Australia and the Antarctic landmass formed by this time.

Table 6.2 Relative time of appearance for species and genera of echinoids in Australia and New Zealand (data taken from Foster & Philip 1978).

	First appearance in Australia	First appearance in New Zealand
Giraliaster (genus)	U. Cretaceous	Eocene
G. sulcatus	lower U. Eocene	middle U. Eocene
G. bellissae	M. Eocene	U. Oligocene
Corystes dysasteroides	U. Eocene	M. Oligocene
Evechinus	U. Miocene	Pliocene
Fellaster	L. Pliocene	Pliocene

6.3.2   Saleniids in time and space

Durkin (1980) provided a detailed account of how the family Saleniidae evolved and how its distribution changed through time. The Saleniidae are a small, well defined group of regular echinoids that first appear in the Upper Jurassic of the Crimea. During the Lower Cretaceous there was a progressive eustatic rise in sea level and the family gradually radiated. Saleniids were restricted to the circum-Mediterranean region in the Valanginian but had expanded into the Gulf province by the Aptian and into northern Europe by the Albian. By the Cenomanian, saleniids were firmly established in shallow-water habitats over a wide geographical area and had reached eastwards to Somalia and western India and southwestwards to Brazil and Angola (Fig. 6.4a).

At the end of the Cenomanian the saleniids suffered a severe setback caused by a major marine regression, and many species became extinct as the shelf seas withdrew, leaving isolated populations. The sea level began

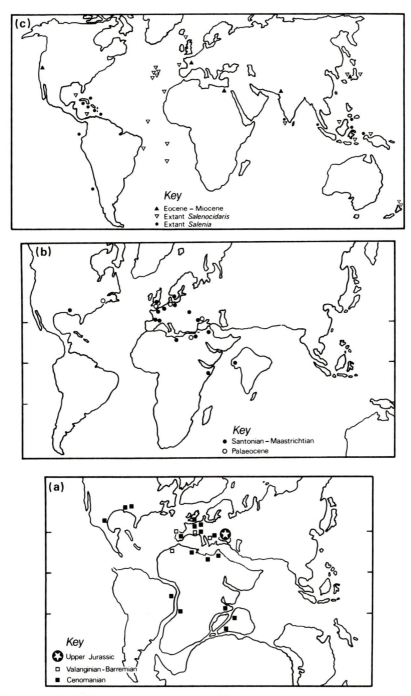

Figure 6.4 Saleniid distribution through time. (a) Upper Jurassic and Lower Cretaceous; (b) Upper Cretaceous and Palaeocene; (c) Eocene to Recent (adapted from Durkin 1979, courtesy of Balkema Press).

to rise again in the Turonian, leading to even more extensive epicontinental seas up to 800 m in depth and extending laterally for thousands of kilometres. Once again, saleniids diversified and spread, mostly adapting to deeper-water facies though with some shallow-water species as, for example, in Sweden (Fig. 6.4b).

A second major regression at the end of the Maastrichtian left isolated populations of saleniids in deep-water basins, notably in northwestern Europe. By the start of the Palaeocene the saleniid fauna had been drastically curtailed. A few species survived for a short time in the deep-water chalk facies of Denmark and the Crimea while others adapted to non-carbonate facies developed in western USA, England and North Africa. Although saleniids once again started to expand during the Tertiary, reaching India and Japan by the Eocene and Australia by the Miocene, they never regained their original widespread distribution (Fig. 6.4c).

Only two genera of saleniid survive today. *Salenia* is found in the Caribbean and Indonesian regions in sublittoral shelf habitats, representing isolated remnants of a once widespread shallow-water genus. The other saleniid, *Salenocidaris*, is found on the mid-Atlantic ridge and adjacent abyssal plains. It apparently evolved from a Cretaceous deep-water population that migrated on to the mid-Atlantic ridge when the Atlantic started to open and has since adapted to live in adjacent deep-sea habitats.

6.3.3   The evolution of sand-dollars

Clypeasteroids provide one of the best documented examples of evolutionary radiation in time and space in echinoderms, due to the work of Durham (1955), Seilacher (1979) and Kier (1982). The oldest known clypeasteroid, *Togocyamus*, comes from the Upper Palaeocene of West Africa, but the group rapidly diversified to become world-wide by the Oligocene and all extant families had evolved by the Miocene. Disc-shaped clypeasteroids, commonly known as sand-dollars, appear to have arisen three times independently during the diversification of this group: once in the Clypeasterina, giving rise to the Arachnoididae, and twice in the Scutellina, giving rise to the Rotulidae and to the superfamily Scutellidea. Each type of sand-dollar has a different geographical centre of origin (Fig 6.5). The Arachnoididae (1) originated in the Australasian region probably when migration first took place from southeastern Africa. The Rotulidae (2) originated in, and remained restricted to, the west coast of Africa. Finally, the Scutellidea (3) originated in the circum-Mediterranean or Gulf and Caribbean region. The precise centre of origin is uncertain since possible 'laganid' ancestors to scutellids are found in the Eocene of France, but the earliest discoidal forms appear in the late Eocene of southern USA. Whereas arachnoidids and rotulids achieved only limited distribution, the scutellids rapidly spread to become almost cosmopolitan. During this initial radiation, the group diversified as it spread out with the

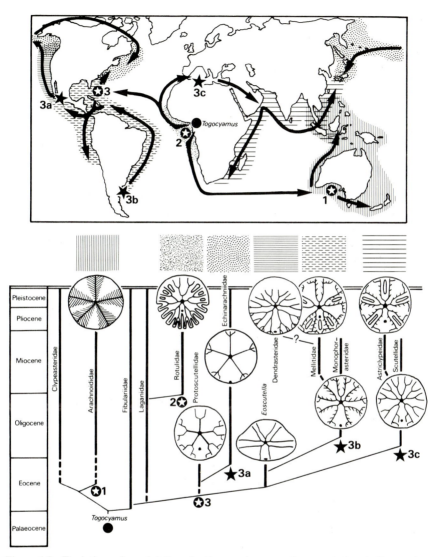

Figure 6.5 Evolution of sand-dollars in time and space. Ornamentation indicates the present-day distribution of the various groups of sand-dollars. Stars and circles mark approximate centres of origin for lineages as indicated; arrows show the suggested migration pathways (highly modified from Seilacher 1979).

result that three independent lines evolved in geographically separated regions (Fig. 6.5):

(a) The Eocene Protoscutellidae of the southeastern USA are too specialised to have given rise to later groups, but are here considered to be the sister group to the Echinarachniidae. Echinarachniids, unlike most other scutellids (apart from protoscutellids) have food grooves that branch well away from the peristome rather than bifurcating on the basicoronal plate. They first appeared in the Oligocene and quickly spread up the west coast of America and around to Japan. By the Pliocene they had also penetrated around the north of Canada to colonise the northwestern Atlantic. *Dendraster*, which originated in the northwestern Pacific in the Pliocene, may have evolved from the echinarachnids as suggested by Durham (1955), but its tooth structure suggests that it is derived from the mellitids (Jensen 1981).

(b) In the western USA the genus *Eoscutella* appeared in the Eocene and appears to have been ancestral to the monophorasterids which diversified in South America during the Oligocene and Miocene. Monophorasterids in turn probably gave rise to the Mellitidae which returned north and successfully displaced the established sand-dollars in northern South America and the Gulf and Caribbean.

(c) The Scutellidae appeared in the circum-Mediterranean region in the Oligocene and diversified to give rise to *Amphiope* and the Astriclypeidae. The astriclpeids rapidly expanded throughout the Indo-Pacific and even reached West Africa during the Miocene. One genus (*Abertella*) successfully colonised the Gulf region in the Miocene but was displaced by the northward-migrating mellitids.

The present-day geographical distribution of sand-dollars is quite easily explained when the evolutionary history of clypeasteroids is examined. In this case, geographical separation brought about by rapid expansion has led to a certain amount of convergence amongst different lineages. The evolution of a discoidal shape which improved their ability to sieve the surface layer of sand for its organic content, the development of branched food grooves which increased the volume of sediment that could be processed at one time, and the evolution of lunules which increased the periphery of the test and hence improved the rate at which food could be harvested, have all occurred independently in different lineages. This parallel evolution was only possible because the ancestral laganinid clypeasteroids had already made the important change to feeding on material that fell between the dorsal spine canopy and which was carried adorally in ciliary currents. This innovation in feeding technique proved highly successful and must have been the principal reason for the dramatic radiation of the clypeasteroids. It was during this radiation that geographically isolated lineages adapted in parallel to cope with similar problems of how to improve their feeding technique.

6.3.4   Origin of the deep-sea fauna

The antiquity of the deep-sea fauna has been discussed widely with suggested ages ranging from 'archaic' to Pliocene. The most recent study (Mironov 1980) concluded that the deep-sea echinoid fauna was Miocene in age and originated in the Indo-West Pacific and Antarctic regions. Some 39 genera in 17 families live at depths greater than 2000 m. These are not primitive unsuccessful groups that have been pushed into the deep-sea environment by more advanced forms, but are often highly specialised groups that have successfully adapted to the rather spartan but predictable deep-sea habitat. Indeed, the deep-sea pourtalesiids are amongst the most specialised of any echinoids.

There is no remnant Palaeozoic element in the deep-sea echinoid fauna contrary to the inference of Philip (1965). All deep-sea echinoid groups can be traced back to shallow-water forms that evolved during the Mesozoic or Tertiary. The fossil record of shallow-water echinoids allows us to estimate the earliest time that individual groups could have invaded the deep-sea environment. For example, deep-water species of the clypeasteroid *Echinocyamus* must be post-Palaeocene arrivals since the genus is not believed to have evolved until the Eocene. Similarly, deep-water species of the regular echinoid *Echinus* are probably even younger. There are groups, notably the echinothurioids and cidaroids, that may possibly have started to move into deeper water as early as the late Triassic, but on the whole most deep-sea groups could only have evolved during the Cretaceous or Tertiary.

More direct evidence comes from trace fossils in deep-sea submarine fan deposits. Crimes (1974) noted that there was a marked increase in the diversity of deep-water traces in the Upper Cretaceous, and this is when the first heart-urchin burrows appear in submarine fan deposits (Smith & Crimes 1983), though they do not become common until the end of the Cretaceous. This suggests that the first major invasion into the deep-sea environment took place towards the end of the Cretaceous. Possibly, prior to this time, there may not have been sufficient organic material reaching the ocean floor for the relatively inefficient detritus-feeding echinoids to survive.

During the Cretaceous there was a global rise in sea level, flooding epicontinental shelf seas to depths of up to 800 m (Hancock & Kauffman 1979). This occurred gradually, allowing echinoids such as *Hagenowia* to adapt to deeper-water conditions and evolve more efficient feeding techniques (see Ch. 5). The major regression that took place at the end of the Cretaceous drastically reduced the area of shelf seas and must have produced intense competition for the available resources. It seems likely that, at this time, several groups that had been adapting to deeper-water habitats simply moved on to the continental slopes and abyssal plains, as was suggested for the saleniid *Salenocidaris* (see p. 134). Since then the deep-sea fauna has continued to diversify during the Tertiary by the migration of offshore shelf species down the continental slope. The only possible example of the reverse situation, where a deep-sea group has given rise to a shallow-water species, is provided by the echinothurioid *Asthenosoma varium* that is found today at depths of less than 100 m in the Indo-Pacific.

6.4   Echinoids as palaeoenvironmental indicators

Since echinoids are often morphologically adapted for living in particular habitats, it should be possible to use them as palaeoenvironmental indicators. The following selected examples give a general idea of the sort of information that can be recovered from fossil echinoids.

6.4.1   Temperature

The evolution of brood pouches has occurred independently in several groups of echinoid. The fact that almost all living marsupiate echinoids are found in Arctic or Antarctic waters suggests that brood pouches evolved as an adaptation for life in cold, harsh environments. Philip and Foster (1971) found that marsupiate echinoids were relatively abundant in the Eocene and Oligocene of Australia but became progressively less common towards the present day when only one species is extant. This they correlate with the rapid migration of the Australian plate northwards, away from Antarctica and into more temperate climes, that took place during the Tertiary.

The relative development of tube feet specialised for gaseous exchange can be another guide to water temperature. Zoeke (1951) first noted that Cretaceous species of the spatangoid *Hemiaster* could be divided into two groups: those from the circum-Mediterranean region had long petals containing many ambulacral pores, whereas those from the North European region had short petals, particularly the posterior pair, and relatively few ambulacral pores (Fig. 6.6). As an echinoid's metabolic rate generally increases with temperature, echinoids living in warmer waters need more

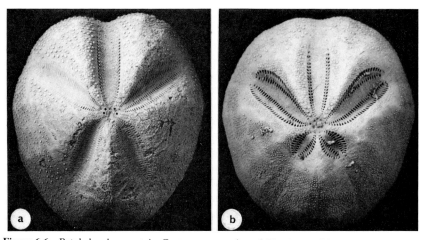

Figure 6.6   Petal development in Cretaceous species of *Hemiaster*. (a) *Hemiaster fourneli* Deshayes from the Senonian of Algeria: a tropical species with elongate petals. (b) *Hemiaster similis* Desor from the Cenomanian of northwestern France: a temperate species with truncated petals. (Photographs courtesy of Porter Kier.)

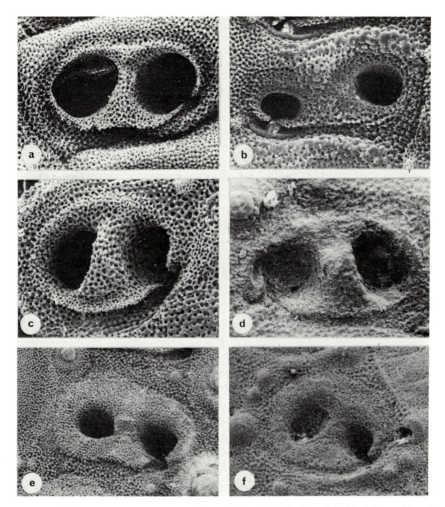

Figure 6.7 Ambulacral pores of Recent and fossil echinoids. (a) Aboral partitioned ambulacral pore of *Poriocidaris purpurata* (Recent), a deep and cold-water cidarid with simple tube feet unspecialised for gaseous exchange (× 80). (b) Aboral conjugate ambulacral pore of *Eucidaris metularia* (Recent), a tropical shallow-water cidarid with flattened and thin-walled tube feet specialised for gaseous exchange (× 80). (c) Adoral ambulacral pore of *Centrostephanus nitidus* (Recent) with a narrow muscle attachment area. This species has weakly muscular tube feet and lives in relatively protected habitats (× 80). (d) Very similar adoral ambulacral pore in the Jurassic pygasteroid *Plesiechinus ornatus* (× 90). (e) Adoral ambulacral pore of *Stomopneustes variolaris* (Recent) with a broad muscle attachment platform. This species has strong muscular tube feet and lives on rocky exposed coasts (× 55). (f) Very similar adoral ambulacral pore in the Jurassic *Stomechinus intermedius*, one of the earliest echinoids to have evolved phyllodes (× 55).

oxygen than those in colder waters. It is therefore no coincidence that species of *Hemiaster* that lived near the Cretaceous palaeoequator had long petals with many tube feet, whereas species in more temperate regions in both hemispheres had short, truncated petals with relatively few tube feet.

A similar type of correlation exists in cidarids (Smith 1978b). Cidarids may either have simple cylindrical tube feet which are associated with partitioned ambulacral pores (Fig. 6.7a) or more complex, flattened tube feet specialised for gaseous exchange, which are associated with conjugate ambulacral pores (Fig. 6.7b). Recent cidarids with these specialised respiratory tube feet are all reef-associated species living in warm, shallow waters. Cidarids from cold-water habitats all have simple cylindrical tube feet.

6.4.2   Turbulence

Adaptations for life in exposed coastal habitats are found in both regular and irregular echinoids. On exposed rocky coasts regular echinoids live in crevices and survive there by clinging tenaciously with their tube feet. Regular echinoids with well developed phyllodes, and oral ambulacral pores with a broad surrounding platform for tube-foot muscle attachment (see p. 41, Fig. 6.7e,f), are adapted for living in highly turbulent habitats. Those without phyllodes or ambulacral pores with large muscle attachment areas (Fig. 6.7c,d) do not have a strong grip and are found in more protected habitats.

Detailed work on living populations of the American sand-dollar *Dendraster* by Raup (1956) and O'Neill (1978) has shown that there is a correlation between test morphology and current velocity. *Dendraster* assumes an inclined posture with only its anterior inserted into the sediment under moderate current regimes (i.e. 100–2000 mm s^{-1}). In this position *Dendraster* feeds by capturing small particles suspended in the water. Raup found that, in populations of *Dendraster* from relativey sheltered bays, individuals had a more centrally positioned apical system than those from open coastal populations, where the apical system lay well back towards the posterior. This, he suggested, was because in more exposed habitats *Dendraster* had to insert more of the test into the sediment for stability. In addition, O'Neill found that bay populations also had tests with higher camber compared to open coastal populations. She attributed this directly to local hydrodynamic conditions, arguing that a test with high camber enhanced lift in low velocity currents and so improved suspension feeding.

Stanton *et al.* (1979) applied Raup's ideas to fossil populations of *Dendraster* from the Pliocene of California. They found that *Dendraster* from the lower part of the stratigraphic section had highly eccentric apical systems, and that eccentricity became less pronounced up the section. This they suggested resulted from the area being progressively cut off from open oceanic conditions through time, producing more sheltered, quieter-water habitats. They also found that within a single formation there was a regional variation in eccentricity, and so they were able to determine the relative exposure and relate this to the palaeogeography.

6.4.3 Substratum

Correlations between morphology and type of sea floor inhabited by the echinoid are relatively common and I shall give just two examples here. As noted above, the regular echinoids with well developed phyllodes and large muscular tube feet are littoral, rocky bottom dwellers. This is a habitat that is rarely preserved in the fossil record as it represents an area of active erosion. However, it is not unusual to find dead tests from these habitats washed into adjacent areas of sedimentation where they stand a better chance of preservation. Therefore the presence of such regular echinoids in fossil communities suggests that shallow rocky bottoms occurred close by. For example, *Stomechinus* and *Pseudopedina* were both adapted for life in exposed rocky habitats, so their presence in the top oolitic beds of the Middle Jurassic Pea Grit Series of the English Cotswolds (see p. 20) suggests that such habitats lay in the vicinity although they are not preserved.

In an extensive study of 51 populations of the spatangoid *Echinocardium cordatum*, Higgins (1975, 1976) identified a number of morphological characters that varied according to the type of sediment that the animals lived in. *E. cordatum* from sands have taller tests, broader plastrons, deeper anterior ambulacral grooves and more funnel-building tube feet in ambulacrum III, than those from mud. Higgins suggested that the difference in the number of funnel-building tube feet might be because it is more difficult to construct and maintain a respiratory shaft in sand than in mud.

7 Classification and phylogeny

There are widely differing views about precisely what makes a good classification, and about the theoretical basis and the methodological approach
that should be used when constructing a classification scheme. Different
approaches not unexpectedly result in quite different groupings and several
alternative classification schemes for the Class Echinoidea are currently
available. Philip (1965), for example, has produced a scheme that groups
echinoids according to their grade of organisation. Durham and Melville
(1957) and Durham (1966) have proposed a different classification based
upon their interpretation of the phylogenetic history, though without distinguishing between groups based on advanced characteristics and those
based on primitive characteristics. A third approach, using cladistic
methodology and recognising only those groups defined on the evolution
of novel characteristics, has been used by Jensen (1981) and Smith (1981).

The classification that I have adopted in this book differs quite strongly
from classifications that will be found in current textbooks and uses a
number of taxonomic groupings that are new. This is because I believe
that, above all, a classification must reflect phylogeny, for reasons which
are well set out by Eldredge and Cracraft (1980). Needless to say,
phylogeny can provide no help in the construction of a classification, since
we have no way of verifying how evolution proceeded but must rely on the
indirect evidence provided by the fossil record and comparative anatomy.
Indeed, the very concept of evolution arose as an attempt to provide a
theory to explain why animals and plants could be classified according to
similarity. If evolution is a branching and diversifying continuum, as we
believe, then homologous morphological similarity amongst organisms has
arisen because of common ancestry. This premise provides the basis for
phylogenetic classification.

During evolution, new morphological features appear and may become
fixed within a species. These new characters will be inherited by all the
descendents of that species except where they become further modified
through evolutionary change. The presence of a morphologically advanced
character in two or more species therefore acts as a marker, identifying those
members that have descended from the common ancestral population
in which the character first became established. A phylogenetic classification seeks to identify **monophyletic** groups (i.e. groups whose mem-
bership comprises all descendants from a common ancestor) by looking
for shared advanced characters (**synapomorphies**). For example the
echinoid keeled tooth represents an evolutionary innovation derived from
grooved teeth. As there is no evidence to suggest that the keeled tooth
evolved more than once, all echinoids with keeled teeth most probably
originate from a common ancestor and therefore form a monophyletic

group (the Echinacea). The alternative state (e.g. the presence of a retained primitive feature or the absence of an advanced character state) does not define a monophyletic group and is termed **plesiomorphic**. For example, some irregular echinoids possess a lantern during part or all of their lives whereas others have lost their lantern completely. The secondary loss of the lantern is an advanced character state and, as it is believed to have occurred only once, defines a monophyletic group (the Atelostomata). The alternative group, composed of echinoids possessing a lantern as juveniles or adults, lumps together a variety of more primitive irregular echinoids because they lack an advanced character. This then defines a **paraphyletic** rather than a monophyletic group.

The higher classification of echinoids has been based on a series of major and stable evolutionary innovations. The main characters used in defining high-level taxonomic categories are the structure of the lantern, the arrangement of ambulacral and interambulacral plates, and the form of the apical and periproctal systems. In addition, pedicellaria, tubercle and spine structure can be helpful and, for particular groups, the arrangement of food grooves and fascioles may also be important.

7.1 Echinoid phylogeny

In the next chapter the pattern of echinoid evolution is analysed and interpreted, but to do this successfully requires a sound phylogeny. A brief

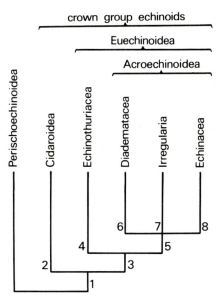

Figure 7.1 Cladogram constructed for major echinoid groups. Points 1–8 are discussed in the text.

outline of the classification and phylogeny of the Echinoidea was given in Chapter 1, and a full scheme to Family level is set out in the Appendix. Some of the evidence for constructing this phylogeny is given here.

Figure 7.1 is a cladogram constructed for the most important echinoid groups. The shared advanced characteristics (synapomorphies) which identify the various monophyletic groups are as follows:

(1) *Crown group Echinoidea.* Crown group echinoids are distinguished from their Palaeozoic stem group Perischoechinoidea because they share the advanced features of having a test composed of just ten ambulacral and ten interambulacral columns, and genital plates perforated by a single gonopore. The earliest crown echinoids belong to the miocidarids and the group arose in the late Permian or Triassic.

(2) *Cidaroidea.* The crown group can be split into two, the Cidaroidea and the Euechinoidea. Cidaroids evolved an upright lantern lacking a foramen magnum, a perignathic girdle of interambulacral apophyses, a joint of the ball and socket type between rotula and epiphyses, solid spines with a cortex, distinctive globiferous pedicellariae, and plates composed largely of rectilinear stereom. All of these features are synapomorphies for the group.

(3) *Euechinoidea.* The synapomorphies that identify this group include the evolution of a perignathic girdle composed of ambulacral auricles, sphaeridia, and external compensation sacs to the peripharyngeal coelom. Euechinoids can themselves be divided into two groups, the Echinothuriacea and the Acroechinoidea.

(4) *Echinothuriacea.* Few advanced features evolved within this group, but one good synapomorphy is the presence of very distinctive ambulacral pseudocompounding.

(5) *Acroechinoidea.* Morphological innovations that unite acroechinoids include the evolution of an upright lantern with a deep, V-shaped foramen magnum, compound ambulacral plating (primitively diadematoid-type compound) and the loss of all but ten of the ambulacral plates and tube feet on the peristome.

(6–8) The Acroechinoidea fall clearly into three groups: two well defined groups and a third plesiomorphic group. Echinacea all share the advanced features of having keeled teeth, solid spines and sutured tessellate plating. The Irregularia, at least initially, all have stout teeth that are diamond- or wedge-shaped in cross section (the complete absence of the lantern is a more recent innovation that has evolved in one subgroup). The third group, Diadematacea, has mostly primitive features as, for example, its retention of grooved teeth (although the tooth structure in living members differs in detail from that of stem group acroechinoids and proves them to be a monophyletic group). Both keeled and diamond-shaped teeth are believed to have evolved independently from grooved teeth, but there is little evidence to show which arose first. The fact that *Eodiadema*, one of the most primitive irregular echinoids, has verticillate spines like those of diadematoids suggests that irregular echinoids and diadematoceans may be more closely related. However, there is really insufficient information about the morphology and evolution of late Triassic and early Jurassic echinoids to be certain of this, and the three groups are left as a trichotomy in the cladogram.

7.2 Classifying Palaeozoic echinoids – the problem of the stem group

The currently accepted classification and phylogeny of Palaeozoic echinoids stems largely from the work of Jackson (1912) and Kier (1965). Many of their groups, however, are paraphyletic, being defined on the presence of primitive features or excluding the most advanced members of the clade. The Lepidocentridae, for example are a heterogenous group whose members are united solely on the fact that they lack advanced features found in other groups. Palaeozoic echinoids pose a particular problem for phylogenetic classification, and this chapter ends by examining the difficulties of integrating the classification of Palaeozoic and post-Palaeozoic echinoids.

Most Palaeozoic echinoids, with the possible exception of some Permian miocidarids, belong to the stem group of the Echinoidea. Precisely the same philosophy can be used to arrange them into a nested heirarchy, as was used on post-Palaeozoic echinoids. Unfortunately, when it comes to giving taxonomic ranks to this hierarchy we are faced with a dilemma. As the ancestry of the crown group is traced progressively backwards through time, each innovation along the way represents a synapomorphy linking all the descendents. Each would identify progressively higher-level taxa and the net result would be an unwieldy, almost meaningless nested series of extremely high level taxa. In an attempt to avoid this problem, Patterson and Rosen (1977) recommended that fossil groups should not be assigned traditional taxonomic ranking but should be designated as **plesions**, monophyletic groups of unspecified rank. The plesion concept provides a neat way of avoiding taxonomic debasement.

The echinoid fauna changed drastically at the end of the Palaeozoic and only one genus, *Miocidaris*, appears to connect the post-Palaeozoic faunas with their Palaeozoic ancestors. It would be rather useful to have a term for members of the stem group and I propose using the term 'Perischoechinoidea' for this, in keeping with McCoy's original concept of the group (McCoy 1849).

It is feasible to construct a cladogram for all genera (Fig. 7.2) because the number of Palaeozoic echinoids is not large. This cladogram can then be translated into a classification scheme (Table 7.1). An integrated

Table 7.1 A classification of Palaeozoic echinoids (see also Fig. 7.2). In this scheme all taxonomic ranks have plesion status and the order in which the genera are listed within higher taxa follows the convention that each is the primitive sister group to all that succeed it. Groupings that are paraphyletic because of the exclusion of *Miocidaris* and all post-Palaeozoic echinoids are placed in inverted commas, and left without formal taxonomic rank. Derived characters on which the cladogram (Fig. 7.2) is based are given in brackets: genera with no characters are plesiomorphic with respect to the preceding higher taxon. Characters indicated by (*) are unique to that group, others are common to all succeeding genera within that group.

SUBCLASS 'Perischoechinoidea'	
Eothuria	(internal radial water vessel; multiperforate ambulacral pores)*
Aulechinus	
Ectinechinus	(loss of ambulacral flexibility and perradial groove)
Aptilechinus	(double ambulacral pores piercing ambulacral plates)
Unnamed paraphyletic group	(pedicellariae; interambulacral spines)
ORDER Echinocystitoida	(ambulacra with more than 2 columns of plates; oligolamellar teeth; primary tubercles on interambulacral plates)
FAMILY Echinocystitidae	
Echinocystites	(interambulacral plates arranged regularly)
Rhenechinus	(interambulacral plates polygonal, tesselate)
Gotlandechinus	(single interambulacral column; only flanges from perradial columns of ambulacral plates enclose radial water vessel)
Cravenechinus	
Xenechinus	
Paraphyletic group 'Palaechinoida'	(interambulacral plates squamose and linearly arranged; apical system of 5 genital and 5 ocular plates; teeth with primary tooth plates)
	(2 poorly known genera with plesiomorphic features)
Incertae sedis	
Koninckiocidaris	(more than 32 interambulacral columns)
Myriastiches	(internal radial water vessel; lepidocentrid-type ambulacral plate imbrication)
SUBORDER Palaeodiscoida	
Palaeodiscus	
FAMILY Lepidocentridae	(interambulacral plates each with eccentrically-positioned primary tubercle)
Lepidechinoides	(internal ambulacral processes found only adorally)*
Lepidocentrus	(ambulacral processes completely lost)
Pholidechinus	(interambulacral plates with large imbrication flange)

SUBFAMILY Proterocidaridae

Pholidocidaris — (ambulacra expanded adorally; oral ambulacral pores large and circular)

Proterocidaris — (large tubercle on adradial interambulacral plates only; ambulacra with about 4 columns)

Pronechinus — (ambulacra broader than interambulacra adorally)

Meekechinus (?) — (ambulacra exceedingly broad adorally and in more than 2 columns adapically)

SUBFAMILY Lepidesthidae

Lepidesthes — (ambulacra in many columns adapically; oral surface unknown)

(ambulacra in many columns throughout)

FAMILY Palaechinidae — (interambulacral plates polygonal, lacking primary tubercles but uniformly covered in miliary tubercles)

Porechinus

Lepidechinus — (plates thick; imbrication flanges small, on ambulacral plates only)

Palaechinus — (basicoronal interambulacral plates resorbed; no imbrication flanges)

Maccoya — (ambulacra in 2 columns, pores biserial)

Lovenechinus — (ambulacra in 4 columns)

Oligoporus ⎤
Melonechinus ⎦ — (ambulacra in more than 4 columns)

Paraphyletic group 'Archaeocidaroida'

(primary tubercle and spine to each interambulacral plate; ambulacral plates with prominent imbrication flange)

Nortonechinus

FAMILY Hyattechinidae — (radial water vessel enclosed only adorally; ambulacral plates enlarged adorally, with large oval pores)

Hyattechinus
Perischodomus

Paraphyletic group 'Archaeocidaridae'

(large hollow primary spines; internal radial water vessel; primordial interambulacral plates resorbed; non-interambulacral peristomial plates)

(every third ambulacral plate enlarged)*

Lepidocidaris
Polytaxocidaris
Archaeocidaris — (interambulacra reduced to 4 columns)

(interambulacra reduced to 2 columns; 1 gonopore per genital plate)

SUBCLASS Cidaroidea
SUBCLASS Euechinoidea

Miocidaris and all post-Palaeozoic echinoids

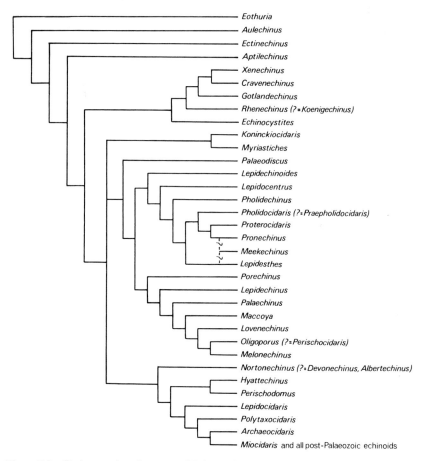

Eothuria
Aulechinus
Ectinechinus
Aptilechinus
Xenechinus
Cravenechinus
Gotlandechinus
Rhenechinus (?= Koenigechinus)
Echinocystites
Koninckiocidaris
Myriastiches
Palaeodiscus
Lepidechinoides
Lepidocentrus
Pholidechinus
Pholidocidaris (?= Praepholidocidaris)
Proterocidaris
Pronechinus
Meekechinus
Lepidesthes
Porechinus
Lepidechinus
Palaechinus
Maccoya
Lovenechinus
Oligoporus (?= Perischocidaris)
Melonechinus
Nortonechinus (?= Devonechinus, Albertechinus)
Hyattechinus
Perischodomus
Lepidocidaris
Polytaxocidaris
Archaeocidaris
Miocidaris and all post-Palaeozoic echinoids

Figure 7.2   Cladogram for all genera of Palaeozoic echinoids. The synapomorphies used to construct this cladogram are listed in Table 7.1.

classification incorporating both Palaeozoic and post-Palaeozoic echinoids raises the problem of taxonomic ranking. Here I have opted to classify the stem group as if it were independent of the crown group. As a consequence, a few groups that are identified are paraphyletic since *Miocidaris* and all post-Palaeozoic echinoids (crown group echinoids) are excluded. The advantage of this system is that a practical classification of the stem group can be formulated, derived directly from the cladogram of character distribution, in which relationships are clearly and unambiguously stated. Furthermore, the system is flexible and can be modified as new evidence becomes available without disrupting the taxonomic hierarchy of post-Palaeozoic echinoids.

8 Evolutionary history

In previous chapters, functional interpretations have been provided for many of the morphological changes that have occurred during the evolution of echinoids and a possible phylogeny has been outlined – it is now time to take a broader look at the pattern of evolution within the group. There is, for example, a marked change in species diversity through time. Echinoids slowly diversified during the Palaeozoic to reach an initial peak in the Lower Carboniferous. There then followed a major decline in species diversity which began in the Upper Carboniferous and continued through the Permian and into the Triassic. Species numbers started to increase again in the Upper Triassic as a major adaptive radiation of echinoids began. This radiation continued through the Mesozoic and much of the Tertiary (Fig. 8.1). Within this broad pattern of changing diversity, the evolutionary history of echinoids can be seen to comprise a series of adaptive breakthroughs that have led to the adoption of new modes of life, or allowed access into new habitats. Each adaptive breakthrough started with the evolution of some crucial new morphological features and led to a period of fairly rapid radiation and morphological diversification as the new niche was filled. Functional explanations can usually be found for morphological innovations and provide possible insight into why particular adaptive thresholds were crossed. This chapter is a largely original attempt to follow how echinoids have evolved to exploit the variety of different habitats in which they are found today.

8.1 Palaeozoic echinoids

The diversity of echinoids during the Palaeozoic appears to have been genuinely low in comparison with post-Palaeozoic echinoids. The low preservational potential of Palaeozoic echinoids must to some extent explain why species diversity seems to have been so low, but the principal cause was probably that Palaeozoic echinoids simply never adapted to live in the wide variety of habitats in which post-Palaeozoic echinoids are found. Throughout the Palaeozoic, echinoids seem to have been broadly restricted to quiet, offshore habitats.

The earliest known echinoids come from the Upper Ordovician. Precisely whence echinoids originated is the subject of the next chapter, but from the functional morphology of these first echinoids it is clear that they were epifaunal deposit feeders. All Cambrian echinoderms were suspension feeders, and it was not until the start of the Ordovician that the first bottom-feeding echinoderms made an appearance. Echinoids evolved as

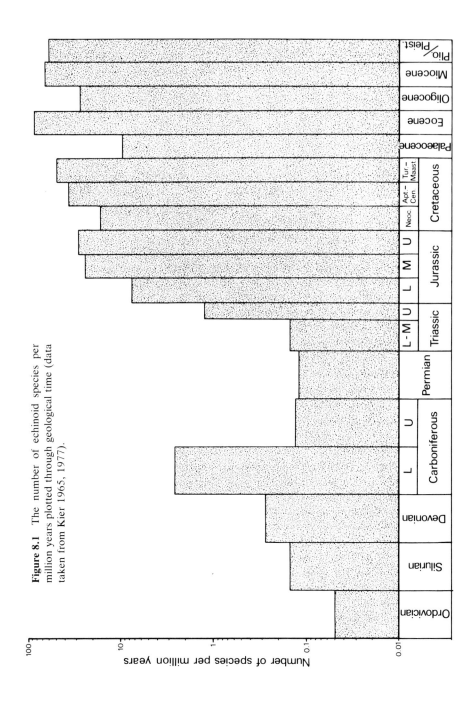

Figure 8.1 The number of echinoid species per million years plotted through geological time (data taken from Kier 1965, 1977).

deposit feeders at a time when echinoderms were experimenting with bottom-feeding techniques. Echinoids must owe their success largely to the evolution of the lantern. Compared with the two other major bottom feeders, starfish and brittle-stars, echinoids possessed a more mobile jaw apparatus, well adapted for cutting, and were probably able to utilise a wider variety of foods, such as algae or organically bound sediment. Although Ordovician echinoids apparently used their lanterns as scoops, by the Lower Silurian a relatively modern-looking lantern had evolved that was clearly used as a biting apparatus. Echinoids therefore made an adaptive breakthrough by evolving a new feeding technique that allowed them to make use of food sources unexploited by other major echinoderm groups.

During the Silurian and Devonian, species numbers slowly increased and echinoids underwent a gradual morphological diversification which established the principal Upper Palaeozoic lineages. Echinoids continued to thrive during the Lower Carboniferous, and there was a marked expansion in the number of species as the newly established lineages diversified. There were four prominent groups during the Carboniferous, each presumably adapted for a different life-style and filling different ecological niches. Proterocidarids and *Hyattechinus*, for example, both evolved flattened tests with enlarged ambulacra on the oral surface. They apparently had large food-gathering oral tube feet and were specialised detritus feeders. Archaeocidarids, in contrast, possessed large defensive spines and were probably fairly opportunistic omnivores, living openly on the sea floor much like the diadematoids of today. Lepidesthids had evolved enormous numbers of tube feet but these were not specialised for gathering food. It is possible that they used their tube feet to hold a covering of material over themselves for camouflage. The fact that both archaeocidarids and lepidesthids appear to have evolved defence strategies suggests that predation (presumably by fish) had become a serious problem during the Devonian. Finally there are the palaechinoids. Just how palaechinoids lived is not yet clear. They are generally associated with interreef sediments and their thick, tesselate tests suggest that they were slow-growing specialists. One strong possibility is that, as they had a dense, uniform covering of tiny spines, they lived hidden amongst loose reef tallus or in nooks and crannies.

The Upper Carboniferous saw a marked decline in echinoid diversity. This was brought about by an overall reduction in species numbers within all existing groups, as well as by extinction of lineages such as the palaechinoids. This trend continued in the Permian where we know of just six species. The successful echinoids at this time fall into two categories. First there were the large, and by then extremely specialised, proterocidarids which had enormous numbers of oral food-gathering tube feet for feeding on detritus. Secondly, there were the small, highly opportunistic omnivores *Miocidaris* and the tiny echinocystid *Xenechinus*. The pronounced decline in echinoid diversity towards the end of the Palaeozoic seems to have occurred because environmental conditions were becoming harsher and less predictable, and there may have been an overall reduction in the primary food supply judging from the extreme specialisation achieved by proterocidarids.

8.2   Post-Palaeozoic regular echinoids

By the Triassic, proterocidarids had become extinct and only the opportunistic generalist *Miocidaris* survived into the Mesozoic and saved the class from extinction. Conditions must at first have remained fairly harsh, for during the Lower and Middle Triassic just three species are known, all of them miocidarids. By the Carnian (early Upper Triassic) however, the situation had changed and echinoid species were once again on the increase. Kier (1977) has reported 16 species from the Carnian. As well as miocidarids, the fauna includes the first true cidaroids and the first acroechinoids as well as the peculiar plesiocidaroids. Slightly later, in the Rhaetic, the first echinaceans are found. Therefore, by the end of the Triassic, miocidarids had undergone a fairly spectacular radiation and the major lineages of post-Palaeozoic regular echinoids had become established.

Both cidaroids and acroechinoids independently evolved stronger lantern and tooth structure, more powerful and efficient lantern muscles attached for the first time to a perignathic girdle, and a sutured test that provided the necessary rigidity for having strong lantern muscles. This suggests that echinoids were adapting to a different diet at this time, one which involved using the lantern with a more powerful plucking or rasping action. The radiation of miocidarids may then have been triggered off by a general improvement in environmental conditions and the appearance of some new source of food which they quickly learned to exploit. The high rate of morphological innovation at this time presumably reflects the lack of competition as echinoids crossed this adaptive threshold and started to diversify rapidly.

The adaptive radiation that began in the Upper Triassic continued well into the Jurassic and it is at this time that the majority of orders of both regular and irregular echinoids first appear. Cidaroids have undergone little change since they appeared in the late Triassic, and have continued to occupy much the same ecological niche. In contrast, acroechinoids have undergone considerable adaptive evolution that has allowed them to colonise previously unexploited habitats. Much of their success must be due to their superior lantern design which is both lighter and more manoeuvrable than cidaroid lanterns. Acroechinoids also developed compound ambulacral plating so as to have larger ambulacral spines and a better all round defence.

At the beginning of the Jurassic, pedinoids and early echinaceans were superficially very similar. However, the mechanically stronger keeled tooth of echinaceans must have given them an advantage over pedinoids, probably by allowing them to feed more efficiently on encrusting organisms. Thus, during the Jurassic, echinaceans were able to expand and diversify as grazers whereas pedinoids remained a fairly conservative group with a less specialised diet.

The rather generalised pseudodiadematids, which are the earliest echinaceans known, gave rise to a variety of different groups during the Jurassic. Members of one such group, the hemicidarids, are convergent with cidarids and presumably evolved to fill a similar niche. Some acrosaleniids also evolved massive interambulacral spines for defence, similar to those of cidarids, but in this group the off-centred periproct suggests that they were experimenting with a diet of sediment. Phymosomatids evolved a short, uniform covering of spines and may have sought protection from predators either by covering themselves or by becoming crevice dwellers. Stomechinids evolved broad phyllodes with strong oral tube feet and were able to colonise shallow rocky habitats which had never before been occupied by echinoids. Their descendants, the arbaciids, became even better adapted for life in very shallow, warm waters by evolving specialised respiratory tube feet.

Diversification of echinaceans continued in the Cretaceous but to a lesser degree. An important improvement came with the evolution of the camarodont lantern in temnopleurids. This is mechanically stronger than the earlier stirodont lantern, and temnopleurids were able to compete successfully with pre-existing stirodont groups. So, by the end of the Cretaceous, temnopleurids were beginning to displace earlier groups, and pseudodiadematids, hemicidarids and acrosaleniids were all extinct. Camarodonts became the dominant regular echinoids during the Tertiary and continued to diversify with the appearance of the Echinoida. These camarodonts were adapted for life in shallow rocky bottoms which brought them into direct competition with stomechinids and arbaciids. Today, stomechinids have declined almost to the point of extinction but the arbaciids have faired slightly better, presumably because of their respiratory tube feet which allow them to live in warmer, shallower water than most other echinoids.

During the evolution of the Echinacea there was an initial diversification of stirodont groups during the Jurassic as they evolved to fill a variety of different niches. This was then followed by a second adaptive radiation that began in the Upper Cretaceous with the evolution of the mechanically superior camarodont lantern. During the course of this radiation, stirodont groups have been progressively displaced from their niches by better adapted camarodont groups.

8.3 Irregular echinoids

The early stages of the evolution of irregular echinoids have been discussed in Chapter 5 and are therefore dealt with only briefly here. Irregular echinoids are believed to have evolved from tiny opportunistic regular echinoids such as *Eodiadema*, which had become adapted to exploit temporarily hospitable areas of the sea floor. They made the adaptive breakthrough to living and feeding on loose mobile sediment early in the Lower Jurassic. Once the threshold had been crossed, diversification progressed rapidly as the group expanded into this major new habitat. Here once again morphological diversification appears to have occurred rapidly at the start of the radiation when there was little competition or selection pressure. The change in life-style from a regular to an irregular echinoid appears to have progressed step-wise. The first irregular echinoids were simply adapted for locomotion over unconsolidated sedimentary bottoms. Later they may have lived semi-infaunally before becoming adapted for burrowing into the sediment, either to avoid being overturned in currents or to hide from potential predators. At about this time they also began to feed on the sediment in which they lived.

The early irregular echinoids, such as pygasteroids and holectypoids, could only burrow into relatively coarse sediments, or lived semi-infaunally. When buried they probably remained more or less stationary and had to return to the surface in order to forage for food, which they still collected using their lantern. However, it was not long before offshoots evolved that were better adapted. Towards the end of the Lower Lias, galeropygoids appeared that were adapted to feed using suckered oral tube feet to pick up their food, and which moved unidirectionally, systematically ploughing through the sediment. Galeropygoids had also evolved a denser, more uniform spine canopy that enabled them to move into finer-grade sediments than were currently being occupied. This adaptive breakthrough was followed by a period of morphological diversification during which the cassiduloid and disasteroid lineages became established.

Initially, disasteroids and cassiduloids appear to have had very similar life-styles and were presumably in direct competition with one another. However, the two groups very quickly became specialised for different niches. Cassiduloids evolved bourrelets and phyllodes and became specialised for swallowing large quantities of sand-sized particles. They were therefore able to colonise areas where the sediments had a relatively low organic content. Disasteroids never adopted feeding techniques whereby large quantities of sediment were ingested. Instead they evolved penicillate tube feet and this allowed them to collect fine-grained material by mucous adhesion. They could therefore be more selective in their feeding and utilise a wider range of particle size in their diet.

The evolution of penicillate tube feet was an important innovation as it apparently resulted in the late Jurassic – early Cretaceous radiation of disasteroids that produced the spatangoid and holasteroid lineages. Quite what went on during this radiation has yet to be worked out in detail. Both spatangoids and holasteroids became very much better adapted for burrowing, and were presumably able to live infaunally within finer or muddier sediments. Spatulate dorsal spines evolved for maintaining the walls of the burrow and in many groups there were also fascioles and subanal tufts of spines. Burrowing in these types of sediment led to the evolution of new feeding techniques and, in some groups, the frontal ambulacrum became an important route for transporting surface detritus down to the mouth. Several lineages have independently adapted for burrowing in muds and other very fine sediments, including at least one group of holasteroids. Generally, however, holasteroids have never been particularly successful burrowers since they lacked the funnel-building tube feet that had evolved in spatangoids. Holasteroids either live epifaunally or burrow in the more permeable sediments where no respiratory shaft is required.

As in most echinoid groups, spatangoid diversity gradually declined towards the end of the Cretaceous and into the Palaeocene (Stokes 1979). Following this decline there was a major radiation in the Eocene during which many of the modern groups arose. Holasteroids never really recovered following the Cretaceous–Tertiary decline, and their former niches were taken over by the spatangoids as they expanded during the Eocene. At present, the only really successful holasteroids are the pourtalesiids, which are specialised for life in the deep-sea environment and have evolved a method of scooping up the surface-detritus layer without using tube feet. In contrast, spatangoids have diversified during the Tertiary and today are found in almost all grades of sediment.

In the early Tertiary an adaptive radiation was also taking place in the clypeasteroids. Clypeasteroids evolved from some form of cassiduloid probably during the Palaeocene. Unlike cassiduloids, they had large numbers of tiny suckered tube feet to each plate and this crucial innovation enabled them to handle finer sediments than cassiduloids and to expand into new habitats. Very quickly clypeasteroids began to use the ciliary currents that brought oxygenated water past the respiratory tube feet, to transport fine detrital particles towards the mouth. This proved so successful that several groups evolved independently to feed solely on these particles, and started sieving fine material from the overlying sediment using their aboral spine canopy as a grille. This change in feeding strategy led to several important morphological adaptations. First of all the test became greatly flattened so that only the topmost layer of organic-rich sediment was sieved. Later the food groove system branched and expanded in order to collect from the entire periphery of the test. Finally the periphery of the test was increased by the development of notches and lunules.

Throughout the evolutionary history of echinoids, new groups have arisen to fill vacant niches or to displace less well adapted groups. Since the Permo-Triassic, when echinoids were within a hair's breadth of becoming extinct, the group has undergone a spectacular radiation. Some lineages have evolved to colonise new habitats but most have evolved to gain access to new sources of food. Today, echinoids probably have a wider diet, and are found in a greater variety of marine habitats than at any time in their past.

9 Origin of the Echinoidea

Echinoids were one of the latest classes of echinoderm to evolve. The oldest undisputed echinoids come from the Upper Ordovician and probably only the holothuroids appeared more recently (*Eldonia* from the Middle Cambrian is not, in my opinion, a holothuroid). From the outset, echinoids are an easily recognisable group without obvious intermediates to other classes. What then was the ancestral echinoderm group that gave rise to the Echinoidea? There is certainly no consensus of opinion about the origin of echinoids and there are about as many theories as there are experts who have written upon the subject. Ancestral groups that have been proposed include helicoplacoids, edrioasteroids, asteroids, ophiuroids, holothuroids, diploporite cystoids and even blastoids.

In this chapter, the problem of echinoid ancestry is tackled first by identifying homologous structures present in other echinoderm groups, and then by identifying morphological innovations that define monophyletic groups using cladistic methodology. Bothriocidaroids, though often claimed to be echinoids (viz. Durham *et al.* 1966), have too many peculiar features to be closely related. Their phylogenetic position is also reviewed below.

9.1 Defining the Class Echinoidea

Before attempting to identify homologous structures within echinoderms, it is worth considering which features are diagnostic for the Class Echinoidea. Although this is an easy task when only living echinoderm classes are compared, it is surprisingly difficult to find any truly diagnostic features when fossil groups are taken into account. Taking just the Upper Ordovician echinoids *Aulechinus*, *Ectinechinus* and *Eothuria* as probable ancestors from which later groups with derived characteristics evolved, then the following diagnosis includes all features which appear to be primitive.

Free-living echinoderms with a globular, flexible test of imbricate plates consisting of a coronal system of five biserial ambulacra and five unorganised interambulacra, and an apical system including five ocular plates and a single genital/madreporite plate situated at the opposite pole to the mouth; a radial water vessel enclosed within ambulacral plates; articulating spines restricted to the adradial margins of ambulacra; a dental apparatus present, composed of modified ambulacral ossicles and fused mouth spines; ambulacral plating around the peristome in obedience to Lovén's Law.

This diagnosis has few features that are unique to echinoids. Other echinoderms of much the same shape include ophiocistioids and holothuroids and there is little that is distinctive about the plating arrangement. Even Lovén's Law, which predicts which of the two columns of plates in each ambulacrum is larger at the peristome, is equally applicable to primitive ophiuroids (Hotchkiss 1979). Moveable appendages are found in asteroids, ophiuroids and possibly edrioasteroids, and a homologous dental apparatus is found in ophiuroids and ophiocistioids. Holothuroids, ophiocistioids and ophiuroids all have enclosed radial water vessels. Thus the Class Echinoidea can only be recognised on the basis of a unique combination of features which individually can be found in other echinoderm groups.

In discussing the origin of echinoids the salient features that have to be explained phylogenetically are: (a) the overall body plan and plating arrangement, (b) the appendages, (c) the jaw apparatus, and (d) the internal position of the radial water vessel.

9.2   Body plan and plating arrangement

Fell (1962, 1967) laid great stress on growth gradients in formulating a higher classification of echinoderms, erecting the Subphylum Asterozoa for star-shaped echinoderms with a radially divergent growth pattern that produces projecting rays, and the Subphylum Echinozoa for essentially globular echinoderms lacking projecting rays and with a meridional growth pattern. This emphasis on body shape is rather superficial (it is hard to imagine how a star-shaped echinoderm could have evolved without radial growth) and masks the detailed homologies that exist amongst eleutherozoan echinoderms which were unravelled in the late 19th and early 20th centuries. Recognising homologies is one of the most important steps in phylogenetic analysis, and Table 9.1 summarises possible homologous structures within various eleutherozoan groups.

The echinoid test consists of two systems of plates: the coronal system, which is homologous with ventral plating in other echinoderms, and the apical system, which is homologous with the entire dorsal surface of other echinoderms. A third series of plates, the peristomial system, is absent in primitive echinoids and evolved within this class.

9.2.1   The coronal system

Pentameral symmetry of ventral plating appears to have been a feature that evolved in echinoderms as early as the Lower Cambrian and has since remained a fairly stable characteristic within eleutherozoans. Ambulacra

Table 9.1 Suggested homologies within eleutherozoan echinoderms.

Stromatocystitoids	Primitive asteroids	Primitive ophiuroids	Echinoids
ventral surface	ventral surface	ventral surface (including arms)	coronal system
biserial flooring plates	ambulacral plates	ambulacral plates	ambulacral plates
primary cover plates	adambulacral ossicles	lateral plates	ambulacral spines
secondary cover plates	adambulacral spines	lateral plate spines	—
interambulacral plates	adaxial plates	ventral disc plating	interambulacral plates
marginal ossicles	inferomarginal ossicles	disc marginals	—
dorsal surface	dorsal surface	dorsal surface (disc only)	apical system
?terminal flooring plates	terminal plates	terminal plates	ocular plates
?	first interradials	first interradials	genital plates
hydropore plate (first posterior interradial)	madreporite	madreporite (? = first precocious interradial)	madreporite
centrodorsal plate	centrodorsal plate	centrodorsal plate	subanal plate
primary oral plates (fused ambulacral plates)	mouth angle plate (= first ambulacral plate)	proximal oral plate (= first ambulacral plate)	hemi-pyramids
	'first ambulacral plate' (= second ambulacral plate)	distal oral plate (= second ambulacral plate)	epiphyses
		?third ambulacral plates	compasses
—	mouth angle spines	?odontophore	rotula
oral cover plates		mouth angle spines	tooth plates
		interradial muscles	interpyramidal muscles
		external radial muscles	intrapyramidal muscles
		1st intervetrebral muscle	protractor/retractor muscles

composed of two alternating series of plates (flooring plates) are a primitive feature for all echinoderms and this arrangement is seen in many groups. The ambulacra of stromatocystitoids, asteroids and early ophiuroids were flexible and could open and close, and a slight degree of flexibility appears to have been retained in the Ordovician echinoid *Aulechinus*. Interambulacral zones of imbricate plating are present in edrioasteroids and in ophiuroid discs. Palaeozoic asteroids generally either lack interambulacral zones or have a regular, serialised pattern of interambulacral plating. Epispires, that are present in interambulacral plates of primitive echinoderms including stromatocystitoids, are not found in echinoids. Internal musculature is still found in living echinoids with imbricate tests and was undoubtedly also present in Palaeozoic echinoids. This corresponds to the dermal layer of circular and longitudinal muscle present in holothuroids, asteroids and euryalid ophiuroids and which was most likely present in stromatocystitoids and edrioasteroids.

9.2.2   The apical system

The apical system of echinoids is homologous with the entire dorsal surface of other eleutherozoans. In Ordovician and Silurian echinoids it consists of five ocular plates, a single genital/madreporite plate and an irregular series of small imbricate plates. Ocular plates are direct homologues of the terminal plates of asteroids and ophiuroids, since both are associated with the tips of radial water vessels and are points of insertion for new ambulacral plates.

The madreporite is a perforated plate formed around the hydropore. A similar plate is present in asteroids, primitive ophiuroids, ophiocistioids and most holothuroids, but is primitively absent in stromatocystitoids and edrioasteroids and has been secondarily lost in post-Palaeozoic ophiuroids. In stromatocystitoids and edrioasteroids the hydropore opens ventrally near the mouth frame through the hydropore oral plate which is probably homologous with the madreporite. An orally positioned hydropore is the primitive condition and has been retained in holothuroids, ophiocistioids and a variety of early asteroids and ophiuroids.

Gonopores are unknown in Palaeozoic asteroids, ophiuroids and edrioasteroids. As the gonopore and hydropore are closely associated in primitive pelmatozoans such as the Lower Cambrian *Kinzercystis*, and as both ophiocistioids and elasipod holothuroids have a single gonopore immediately adjacent to the hydropore, it seems likely that edrioasteroids and stromatocystitoids also had a single gonad that opened directly into the hydropore. Primitive echinoids were therefore similar to other early echinoderms in having a single gonad and a gonopore that opened in close association with the hydropore.

The embryological development of the skeleton is similar in echinoids and asteroids. Initially both have 11 aboral plates; five radial pieces (oculars/terminal plates), five interradial plates (the genital plates of echinoids), one of which grows around the hydropore, and a single centrodorsal plate. The dorsal skeleton in ophiuroids likewise starts with 11 plates, but as well as five terminal plates and a centrodorsal plate there are a further five radially positioned plates, possibly equivalent to the radially interpolated plates that appear in asteroids. Interradial plates, of which the earliest to form is possibly homologous with the madreporite (Hendler 1978), appear slightly later embryologically. Hendler rejected any close homology between the initial skeletal plating of ophiuroids and echinoids since these formed from larval spicules in echinoids, but arose independently of larval spicules in ophiuroids. However, as only two of the ocular plates and three of the genital plates arise from the larval skeleton of echinoids (Hyman 1955), this difference does not seem too important.

In their body plan, echinoids show many similarities with other eleutherozoan groups and, in this respect, there is no fundamental difference between echinoids and asterozoans. A globular-bodied echinoderm could easily have been derived from an asteroid or ophiuroid simply by retaining the terminal plates apically during embryological development while continuing to add new plates ventrally.

9.3   Appendages

The only articulating appendages found in Upper Ordovician and Lower Silurian echinoids are spines; pedicellariae and sphaeridia being later innovations. In these early echinoids, spines are restricted to the adradial margins of ambulacral plates and only later did they develop in interambulacral areas. These initial ambulacral spines appear to be homologous with cover plates in stromatocystitoids and edrioasteroids. Bather (1900) suggested that the ambulacral plates of echinoids were homologous with the cover plates of edrioasteroids, based on the position of the radial water vessel, but this does not match palaeontological evidence. Stromatocystitoids had a skirt of cover plates attached to the adradial edge of the flooring plates (Fig. 9.1). The first series of cover plates (primary cover plates) were larger than the rest and the secondary series attached directly to them. The primary cover plates later became transformed into adambulacral ossicles in asteroids and lateral arm plates in ophiuroids, but continued to serve the same function of protecting the radial water vessel. In early ophiuroids and asteroids, these plates usually remained moveable and articulated near the adradial edge of the flooring plates (Spencer 1914–40). Secondary cover plates are presumably homologous with the spines carried by adambulacral ossicles and lateral arm plates. In echinoids, primary and possibly secondary cover plates were reduced to adambulacral spines which continued to protect the tube feet and perradial groove.

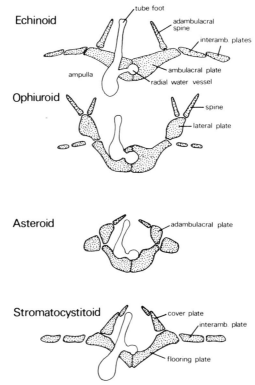

Figure 9.1 Diagrammatic cross sections through ambulacra of primitive eleutherozoan echinoderms to illustrate possible homologies in plating. External surface is uppermost in all diagrams.

9.4 Jaw apparatus

The lantern is one of the most distinctive features of echinoids, yet an almost identical system is present in ophiocistioids (Haude & Langenstrassen 1976, Fig. 9.2). The homologies between the echinoid and ophiuroid jaw apparatuses have been oulined by Devanesen (1922). Both consist of modified ambulacral ossicles with the first pair from adjacent ambulacra connected interradially (Fig. 9.2). Echinoid hemi-pyramids are homologous with the proximal oral plates (first ambulacral plates according to Hendler 1978) of ophiuroids, while the paired epiphyses lying on either side of the radial water vessel are homologous with the distal oral plates (second ambulacral plates). The teeth, which in *Aulechinus* are no more than a series of spines welded together, are homologous with mouth spines in ophiuroids and asteroids.

The origin of the rotulae poses more of a problem, since they are unpaired radial structures with no obvious homologue in asterozoans. Devanesen (1922) suggested that the ophiuroid odontophore might be

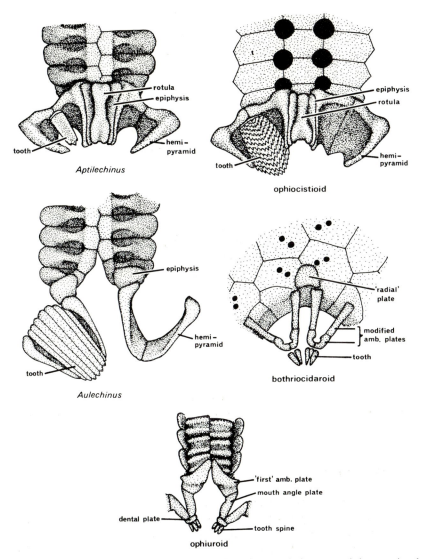

Figure 9.2 Reconstructions of the proximal plating of one ambulacrum and the associated jaw apparatus in primitive eleutherozoan echinoderms viewed from the interior of the test. *Aulechinus* is an Ordovician echinoid, *Aptilechinus* a Lower Silurian echinoid. The reconstruction of the bothriocidaroid jaw is based on the description by Mannil (1962), and that of the ophiocistioid jaw on the description by Haude and Langenstrassen (1976).

homologous but this seems unlikely, for although the odontophore is an unpaired element, it is interradial in position. Rotulae may then be an evolutionary innovation that has arisen in the ancestors to echinoids and ophiocistioids. Compasses, possibly originating from a modified third pair of ambulacral ossicles, evolved after the Ordovician within the Echinoidea.

Finally musculature in echinoid, ophiuroid and presumably ophiocistioid jaw apparatuses can also be homologised (see Table 9.1).

Jaw homologies can be traced back into earlier groups. Somasteroids and some asteroids have mouth frames consisting of ambulacral plates that are joined interradially and the same is true in *Stromatocystites* from the Lower Cambrian, though here a posterior interambulacral plate (the hydropore oral plate) is also inserted. Other asteroids have mouth angle plates of adambulacral origin which represents a derived state.

The jaw apparatus in the Ordovician echinoid *Ectinechinus* (Fig. 9.2) bears some resemblance to contemporary ophiuroid jaw apparatuses and both were probably equally mobile.

9.5 The water vascular system

Echinoids have radial water vessels that lie internal to ambulacral plates. During larval development the radial water vessels are initially external but later on become closed off by the formation of epineural folds. This has long been taken as evidence that echinoids evolved from ancestors with an external system of radial water vessels, and agrees with the fact that in primitive echinoids the radial water vessels were not fully internal but lay enclosed within the ambulacral plates (Fig. 9.1).

Amongst other eleutherozoan echinoderms a comparable arrangement is found in Lower Palaeozoic ophiuroids where the radial water vessel is partially or completely roofed over by extensions of the ambulacral plates (Fig. 9.1). In the embryological development of living ophiuroids, the radial water vessel becomes enclosed precisely as in echinoids, and the same is also true for holothuroids. Asteroids are more primitive in having external radial water vessels, and this was almost certainly also the case in stromatocystitoids. Echinoids, ophiuroids, holothuroids, and most probably ophiocistioids and bothriocidarids, all share the advanced character of having an enclosed radial water vessel.

Side branches from the radial vessels give rise to tube feet which, in echinoids, are connected to internal ampullae. Tube feet in all living eleutherozoans are remarkably similar and differ from those of crinoids (Nichols 1972). Further differences between crinoids and eleutherozoan echinoderms are found in the arrangement of stone canals and hydropores. In asteroids, echinoids and possibly in early ophiuroids the stone canal in cross section has the shape of a double scroll.

9.6 Other evidence

In addition to the clues about the origin of echinoids that can be obtained from tracing the evolution of homologous structures, other lines of evidence are available. These come from studies of the biochemistry and embryology of living echinoderms.

9.6.1 Biochemical evidence

Studies on biochemical similarity amongst echinoderms have been carried out using sterols (Bolker 1967), phosphorus carriers (Florkin 1952) and collagen (Matsumura *et al.* 1979). These show that there is strong biochemical similarity between ophiuroids and echinoids. Both have Δ5 sterols (cholesterols) and creatine as a phosphorus carrier, whereas asteroids and holothuroids have Δ7 sterols (stellasterols) and arginine as a phosphorus carrier. Analysis of collagen biochemistry gives a similar result.

9.6.2 Larval development

Fell (1945, 1967) rejected the idea that the larvae of echinoids provide any evidence of their phylogenetic relationships. His arguments were based on the occurrence of modified ontogenies that are developed in certain groups. Although these are often striking, they are specialised modifications foreshortening the more usual (and more primitive) development. These have little relevance, as Hyman (1955) pointed out, since phylogenetic arguments should be based on the full development found in each group.

When this is done, it is clear that the pluteus larvae of echinoids and ophiuroids, though not identical, are strikingly similar. Both pass through an early (dipleurula) stage which is also present in asteroids and holothuroids, but later larval stages are quite different from those in other echinoderms (Fig. 9.3). This larval similarity was explained as convergence by Fell (1967) largely because the fossil record shows that asteroids and ophiuroids share a common ancestor. As these two groups divided in the early Ordovician, a more plausible explanation is that echinoids evolved from primitive ophiuroids after they had split off from asteroids.

9.7 Echinoid ancestry

In tracing the evolution of homologous structures within eleutherozoan echinoderms, it becomes clear that ophiuroids and echinoids share a number of important evolutionary innovations (Fig. 9.4), indicating that they separated relatively recently. The fossil record gives fairly convincing evidence of a gradual transformation from a stem asterozoan such as *Archegonaster* to present-day ophiuroids. The results of cladistic analysis imply that echinoids evolved from a group of primitive ophiuroids, after ophiuroids had split off from asteroids but before they had evolved diametrically opposing ambulacral plating.

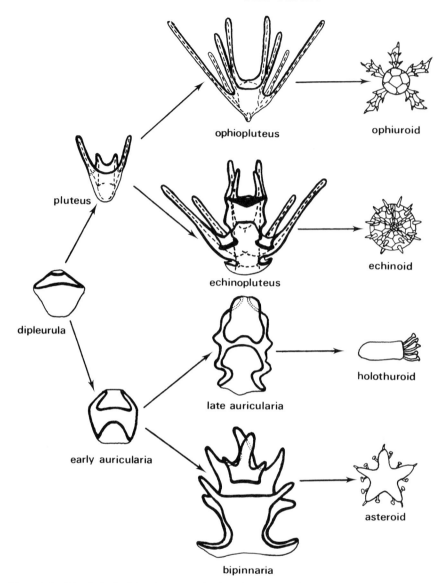

Figure 9.3 Morphological development of echinoderm larvae (taken from Fell 1967, courtesy of the Geological Society of America).

The transformation from some form of primitive ophiuroid into a primitive echinoid is surprisingly easy. Three basic steps are required:

(a) During embryological development the terminal plates must be retained apically while continuing to initiate oral (coronal) plating as in ophiuroids. The first interradial (madreporite) must also be retained aborally at this stage.

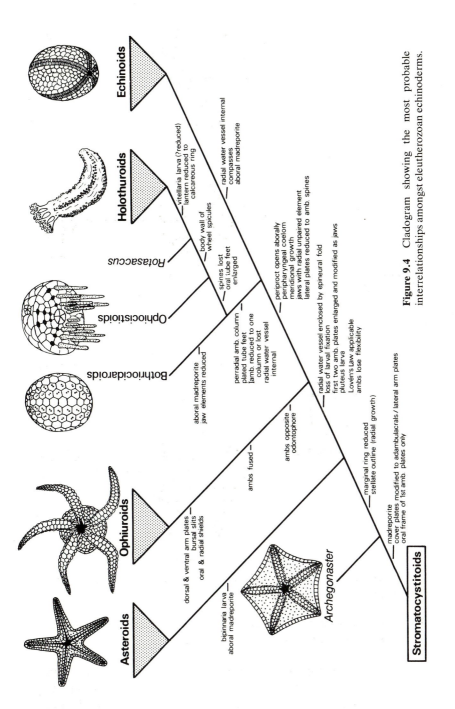

Figure 9.4 Cladogram showing the most probable interrelationships amongst eleutherozoan echinoderms.

(b) The lateral arm plates (cover plates) must be reduced to ambulacral spines but continue to serve the same protective function.

(c) The mouth frame plates must develop further and become more mobile (this took place largely within the Echinoidea) and a new unpaired radial ossicle, the rotula, must form.

The ancestral ophiuroid should therefore have had slightly flexible, alternating ambulacral plating, an enclosed radial water vessel, a well developed jaw apparatus, lateral plates reduced on the disc and a single marginally positioned madreporite. This points to a member of the Proturina, Parophiurina or Lysophiurina as being possibly ancestral to the Echinoidea.

9.8 The position of *Bothriocidaris*

The Order Bothriocidaroida contains three genera (*Bothriocidaris* (Fig. 9.5), *Neobothriocidaris* and *Unibothriocidaris*) which range from the Middle Ordovician (Llanvirnian) to the Upper Silurian (Ludlovian). These animals have usually been considered to be primitive echinoids (e.g. Durham 1966) and antedate the earliest unquestionable fossil echinoid. Bothriocidaroids have been classified as echinoids because they possess a lantern apparatus, articulating spines, an internal radial water vessel and an arrangement of oral ambulacral plates that obeys Lovén's Law (though this last point has been questioned by Paul 1967). They also have meridional

Figure 9.5 *Bothriocidaris pahleni* (Ordovician); lateral view, × 6. Plates of the apical system can be seen at the top and two columns of perforate plates with a single column of imperforate plates on either side are clearly seen. A few spines remain adhering to the test.

growth, and dorsal plating that is reduced to a small apical system and includes a madreporite and four other interradial plates.

Bothriocidaroids are therefore quite obviously more closely related to echinoids than to ophiuroids, but they have several pecular features which suggest that they lie off the main ophiuroid–echinoid lineage. The most important of these are: (a) the presence of a uniserial row of imperforate radial plates overlying or sometimes enclosing the radial water vessel and separating the adradial series of 'ambulacral' plates; (b) the reduction of interambulacral areas to a uniserial row, or their complete loss; (c) the heavily calcified tube feet; (d) tesselate plating. These are all advanced features not found in either echinoids or ophiuroids. They are, however, characteristic features of ophiocistioids.

The 'lantern' in *Bothriocidaris* is very peculiar, as noted by Mannil (1962). There are two paired (?ambulacral) pieces, an unpaired radial plate and tiny mouth spines (Fig. 9.2), and so the bothriocidaroid lantern can be homologised with those of echinoids and ophiocistioids. However, these lantern plates are small, fragile and poorly developed, and in my opinion probably represent a degenerate, non-functioning apparatus that has been retained in a reduced form from an ancestor with a more fully developed lantern.

There is good reason to believe that holothuroids are the most closely related living group to echinoids, and that they evolved from ophiocistioids through the Devonian genus *Rotasaccus*. One puzzling feature of holothuroids is that they have a large peripharyngeal coelom with an associated internal series of radial and interradial plates, known as the calcareous ring. As a peripharyngeal coelom is otherwise only found in echinoids where it encloses the lantern, this has been taken as evidence that holothuroids evolved from ancestors with a well developed lantern, and that the calcareous ring derives from lantern elements (MacBride 1906). Haude and Langenstrassen (1976) described a very peculiar ophiocistioid-like animal *Rotasaccus* that has advanced holothuroid features. *Rotasaccus* has a large lantern and presumably a peripharyngeal coelom, but its body wall skeleton is reduced to tiny wheel-shaped spicules, a feature otherwise unique to holothuroids.

Since ophiocistioids and bothriocidaroids have features that suggest they are sister groups, and since holothuroids possibly evolved from ophiocistioids, bothriocidaroids are best considered in phylogenetic terms as stem holothuroids rather than as echinoids (Fig. 9.4).

Appendix

Classification and stratigraphical ranges of echinoid families

Class Echinoidea Leske, 1778
  *Subclass Perischoechinoidea McCoy, 1849 (see Table 7.1 for details)
  Subclass Cidaroidea Claus, 1880 (*nom. transl.*)
    Order Cidaroida Claus, 1880
      Family Cidaridae Gray, 1825
        Psychocidaridae Ikeda, 1938
        *   Diplocidaridae Gregory, 1900
  Subclass Euechinoidea Bronn, 1860
  Infraclass Echinothurioidea Claus, 1880 (*nom. transl.*)
    Cohort Echinothuriacea Jensen, 1981
      Order Echinothurioida Claus, 1880
        Family Echinothuriidae Thomson, 1872
          Phormosomatidae Jensen, 1981
        *   Pelanechinidae Groom, 1887
  Infraclass Acroechinoidea Smith, 1981
    Cohort Diadematacea Duncan, 1889
      Order Diadematoida Duncan, 1889
        Family Diadematidae Gray, 1855
          Aspidodiadematidae Duncan, 1889
          Lisodiadematidae Fell, 1966
      Order Micropygoida Jensen, 1981
        Family Micropygidae Jensen, 1981
      Order Pedinoida Mortensen, 1939
        Family Pedinidae Pomel, 1883
    Cohort Echinacea Claus, 1876
        *Family Pseudodiadematidae Pomel, 1883
      Superorder Stirodonta Jackson, 1912
        *Family Hemicidaridae Wright, 1857
      Order Salenioida Delage and Herouard, 1903
        *Family Acrosaleniidae Gregory, 1900
          Saleniidae L. Agassiz, 1838
      Order Phymosomatoida Mortensen, 1904
        Family Phymosomatidae Pomel, 1883
        *   Glyptocidaridae Jensen, 1981
        *   Stomechinidae Pomel, 1883
          Stomopneustidae Mortensen, 1903
          Arbaciidae Gray, 1855
      Superorder Camarodonta Jackson, 1912
        *Family Glyphocyphidae Duncan, 1889
      Order Temnopleuroida Mortensen, 1941
        Family Temnopleuridae A. Agassiz, 1872
      Order Echinoida Claus, 1876

Family Echinidae Gray, 1825
Echinometridae Gray, 1825
Parechinidae Jensen, 1981
Strongylocentrotidae Gregory, 1900
Toxopneustidae Troschel, 1872
Cohort Irregularia Latreille, 1825
*Family Eodiadematidae (nov.)
Superorder Eognathostomata Smith, 1981
*Order Pygasteroida Durham and Melville, 1957
*Family Pygasteridae Lambert, 1900
Order Holectypoida Duncan, 1889
*Suborder Holectypina Duncan, 1889
*Family Holectypidae Lambert, 1899
*   Anorthopygidae Wagner and Durham, 1966
*   Discoididae Lambert, 1899
Suborder Echinoneina H. L. Clark, 1925
Family Echinoneidae Agassiz and Desor, 1847
*   Conulidae Lambert, 1911
*   Galeritidae Gray, 1825
*   Neoglobatoridae Endelman, 1980
Superorder Microstomata (nov.)
Series Neognathostomata Smith, 1981
*Family Galeropygidae Lambert, 1911
Order Cassiduloida Claus, 1880
Family Cassidulidae Agassiz and Desor, 1847
*   Archiacidae Cotteau and Triger, 1869
*   Clypeidae Lambert, 1898
*   Clypeolampadidae Kier, 1962
*   Conoclypeidae Zittel, 1879
Echinolampadidae Gray, 1851
*   Faujasiidae Lambert, 1905
Nucleolitidae Agassiz and Desor, 1847
Pliolampadidae Kier, 1962
*Order Oligopygoida Kier, 1967
*Family Oligopygidae Duncan, 1889
Order Clypeasteroida A. Agassiz, 1872
Suborder Clypeasterina A. Agassiz, 1872
Family Clypeasteridae L. Agassiz, 1835
Arachnoididae Duncan, 1889
Suborder Scutellina Haeckel, 1896 (emend.)
Infraorder Fibulariina (nov.)
Family Fibulariidae Gray, 1825
Infraorder Laganina Mortensen 1948 (emend.)
Superfamily Laganidea A. Agassiz, 1873 (*nom. transl.*)
Family Laganidae A. Agassiz, 1873
Rotulidae Gray, 1855
*   Neolaganidae Durham, 1954
Superfamily Scutellidea (nov.)
*Family Scutellidae Gray, 1825

                              Astriclypeidae Stefanini, 1911
                              Dendrasteridae Lambert, 1889
                              Echinarachniidae Lambert, 1914
                              Mellitidae Stefanini, 1911
              *   Monophorasteridae Lahille, 1896
              *   Protoscutellidae Durham, 1955
          Order Neolampadoida Philips, 1963
                      Family Neolampadidae Lambert, 1918
      Series Atelostomata Zittel, 1879
      *Order Disasteroida Mintz, 1968
                  *Family Disasteridae Gras, 1848
                  *Family Collyritidae d'Orbigny, 1853
          Order Holasteroida Durham and Melville, 1954
                  *Family Holasteridae Pictet, 1857
                      Calymnidae Mortensen, 1907
              *   Somaliasteridae Wagner and Durham, 1966
              *   Stenonasteridae Lambert, 1922
                      Urechinidae Duncan, 1889
                      Pourtalesiidae A. Agassiz, 1881
          Order Spatangoida Claus, 1876
                  Family Toxasteridae Lambert, 1920
          Suborder Micrasterina Fischer, 1966
                  *Family Micrasteridae Lambert, 1920
                      Brissidae Gray, 1855
                      Loveniidae Lambert, 1905
                      Spatangidae Gray, 1825
          Suborder Hemiasterina Fischer, 1966
                  Family Hemiasteridae Clark, 1917
                      Palaeostomatidae Loven, 1867
                      Aeropsidae Lambert, 1896
                      Schizasteridae Lambert, 1905
                      Pericosmidae Lambert, 1905
          Suborder unknown
                  Family Asterostomatidae Pictet, 1857

    The following groups are unplaced:
      *Order Plesiocidaroida Duncan, 1889
        *Family Tiarechinidae Gregory, 1896 (subclass uncertain)
      *Order Orthopsida Mortensen, 1942
        *Family Orthopsidae Duncan, 1889 (Cohort Diadematacea or Echinacea)

    * Extinct groups.

Diagnoses for new taxa

Family Eodiadematidae

Periproct lying within apical system: all genital plates perforate: primary tubercles
perforate, crenulate: spines hollow: lantern present in adults: teeth diamond-shaped

in cross section, with broad triangular side plates that overlap and a reduced prism zone. Type genus *Eodiadema*.

Superfamily Scutellidea

Clypeasteroids with a flattened test and a system of branched food grooves: interambulacral plates in two columns apically: lantern support structures fused auricles: aboral miliary spines ending in a glandular bag: two spicules in disc of suckered tube feet. Type genus *Scutella*.

Infraorder Fibulariina

Small clypeasteroids with perignathic girdle of fused auricles: pores on oral surface arranged in discrete bands: (petals simple: internal supports absent or simple radial partitions: food groove system absent). Type genus *Fibularia*.

Infraorder Laganina Mortensen (emend.)

Generally flattened clypeasteroids with petalloid ambulacra and simple or branched food grooves: radial and concentric internal supports or complex pillering: periproct on oral surface: interambulacral areas narrow: perignathic girdle of fused auricles. Type genus *Laganum*.

Superorder Microstomata

Primordial ambulacral plates incorporated into the corona. No ambulacral plate compounding: no primary tubercles: tubercles perforate, crenulate: spines hollow: tubercles on oral surface arranged for unidirectional locomotion: peristome small, without buccal slits: lantern present or absent in adults, wholly internal and unable to protrude through peristome: teeth diamond- or wedge-shaped in cross section when present. Type genus *Cassidulus*.

References

Alexander, D. E. and J. Ghiold 1980. The functional significance of the lunules in the sand dollar *Mellita quinquiesperforata*. *Biol Bull.* **159**, 561–70.

Aslin, C. J. 1968. Echinoid preservation in the Upper Estuarine Limestone of Bilsworth, Northamptonshire. *Geol Mag.* **105**, 506–18.

Bather, F. A. 1900. The echinoderms. In *A treatise on zoology*, vol. III, E. R. Lankester (ed.), 1–344. London: A. and C. Black.

Birkeland, C. and F. S. Chia 1971. Recruitment risk, growth, age and predation in two populations of sand dollars, *Dendraster excentricus* (Eschscholtz). *J. Exp. Mar. Biol. Ecol.* **6**, 265–78.

Blake, D. B. 1968. Pedicellariae of two Silurian echinoids from western England. *Palaeont.* **11**, 576–9.

Bloos, G. 1973. Ein Fund von Seeigeln der Gattung *Diademopsis* aus dem Hettangium Württembergs und ihr Lebensraum. *Stuttgarter Beitr. Naturk. (B)* **5**, 1–25.

Bolker, H. I. 1967. Phylogenetic relationships of echinoderms: biochemical evidence. *Nature* **213**, 904–5.

Boolootian, R. A. (ed.) 1966. *Physiology of Echinodermata*. London: Wiley Interscience.

Bromley, R. G. and U. Asgaard 1975. Sediment structures produced by a spatangoid echinoid: a problem of preservation. *Bull. Geol Soc. Denm.* **24**, 261–81.

Campbell, A. C. 1973. Observations on the activity of echinoid pedicellariae, I. Stem responses and their significance. *Mar. Behav. Physiol.* **2**, 33–61.

Campbell, A. C. 1974. Observations on the activity of echinoid pedicellariae, II. Jaw responses of tridentate and ophicephalous pedicellariae. *Mar. Behav. Physiol.* **3**, 17–34.

Campbell A. C. 1976. Observations on the activity of echinoid pedicellariae, III. Jaw response of globiferous pedicellariae and their significance. *Mar. Behav. Physiol.* **4**, 25–39.

Campbell, A. C., J. K. G. Dart, S. M. Head and R. F. G. Ormond 1973. The feeding activity of *Echinostrephus molaris* (de Blainville) in the central Red Sea. *Mar. Behav. Physiol.* **2**, 155–69.

Chesher, R. H. 1966. The R/V Pillsbury Deep-Sea Biological Expedition to the Gulf of Guinea, 1964–65. 10, Report on the Echinoidea collected by R/V Pillsbury in the Gulf of Guinea. *Stud. Trop. Oceanogr. Miami* **4**, 209–23.

Chesher, R. H. 1968. The systematics of sympatric species in West Indian spatangoids: a revision of the genera *Brissopsis*, *Plethotaenia*, *Paleopneustes* and *Saviniaster*. *Stud. Trop. Oceanogr. Miami* **7**, 1–168.

Chesher, R. H. 1970. Evolution of the genus *Meoma* (Echinoidea: Spatangoida) and a description of a new species from Panama. *Bull. Mar. Sci.* **20**, 731–61.

Clarke, M. W. H. and A. J. Keij 1973. Organisms as producers of carbonate sediments and indicators of environments in the southern Persian Gulf. In *The Persian Gulf: Holocene carbonate sedimentation and diagenesis in a shallow epicontinental sea*, B. H. Purser (ed.), 33–56. Berlin: Springer.

Crapp, G. B. and M. E. Willis 1975. Age determination in the sea urchin *Paracentrotus lividus* (Lamarck) with notes on the reproductive cycle. *J. Exp. Mar. Biol. Ecol.* **20**, 157–78.

Crimes, T. P. 1974. Colonisation of the early ocean floor. *Nature* **248**, 328–30.

Currey, J. D. and D. Nichols 1967. Absence of an organic phase in echinoderm calcite. *Nature* **214**, 81–3.

Davies, T. T., M. A. Crenshaw and B. M. Heatfield 1972. The effect of temperature on the chemistry and structure of echinoid spine regeneration. *J. Paleont.* **46**, 874–83.

Deutler, F. 1926. Über das Wachstum des Seeigelskeletts. *Zool. Jb. Abt. Anat. Ontog. Tier.* **48**, 119–200.

Devanesen, D. W. 1922. Development of the calcareous parts of the lantern of Aristotle in *Echinus miliaris. Proc. R. Soc. Lond. (B)* **93**, 468–92.

Durham, J. W. 1955. Classification of clypeasteroid echinoids. *California Univ. Publ. Geol Sci.* **31**, 73–198.

Durham, J. W. 1966. Evolution among the Echinoidea. *Biol Rev.* **41**, 368–91.

Durham J. W. and R. V. Melville 1957. A classification of echinoids. *J. Paleont.* **31**, 242–72.

Durham, J. W., H. B. Fell, A. G. Fischer, P. M. Kier, R. V. Melville, D. L. Pawson and C. D. Wagner 1966. Echinoids. In *Treatise on invertebrate palaeontology, part U: Echinodermata 3*, R. C. Moore (ed.). Lawrence, Kansas: University of Kansas Press and the Geological Society of America.

Durkin, M. K. 1980. The Saleniidae in time and space. In *Echinoderms: present and past*, M. Jangoux (ed.). 3–14. Rotterdam: Balkema.

Ebert, T. A. 1967. Negative growth and longevity in the purple sea-urchin *Stronglyocentrotus purpuratus* (Stimpson). *Science* **157**, 557–8.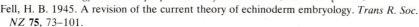

Ebert, T. A. 1971. A preliminary quantitative survey of the echinoid fauna of Kealakekua and Honaunau Bays, Hawaii. *Pac. Sci.* **25**, 112–31.

Ebert, T. A. 1975. Growth and mortality of post-larval echinoids. *Am. Zool.* **15**, 755–75.

Ebert, T. A. 1982. Longevity, life history, and relative body wall size in sea urchins. *Ecol Monogr.* **52**, 353–94.

Eldredge, N. and J. Cracraft 1980. *Phylogenetic patterns and the evolutionary process*. New York: Columbia University Press.

Eldredge, N. and S. J. Gould 1972. Punctuated equilibrium: an alternative to phyletic gradualism. In *Models in paleobiology*, T. J. M. Schopf (ed.), 82–115. San Francisco: Freeman Cooper.

Ernst, G. 1972. Grundfragen der Stammesgeschichte bei irregulären Echinden der nordwesteuropäischen Oberkreide. *Geol. Jb. (A)* **4**, 63–175.

Ernst, G. 1973a. Aktuopaläontologie und Merkmalsvariabilität bei mediterranen Echiniden und Rückschlüsse auf die Ökologie und Artumgrenzung fossiler Formen. *Paläont. Z.* **47**, 188–216.

Ernst, G. 1973b. Die Echiniden-Fauna aus dem Santon der Gehrdener Berge. *Berichte Natur. Ges.* **117**, 79–102.

Farmanfarmaian, A. and A. C. Giese 1963. Thermal tolerance and acclimation in the purple sea urchin *Strongylocentrotus purpuratus. Physiol. Zool.* **36**, 237–43.

Fell, H. B. 1945. A revision of the current theory of echinoderm embryology. *Trans R. Soc. NZ* **75**, 73–101.

Fell, H. B. 1962. A classification of echinoderms. *Tuatara* **10**, 138–40.

Fell, H. B. 1967. Echinoderm ontogeny. In *Treatise on invertebrate paleontology, part S: echinodermata 1*, R. C. Moore (ed.), 60–85. Lawrence, Kansas: University of Kansas Press and the Geological Society of America.

Florkin, M. 1952. Caractères biochimique des categories supraspecifiques de la systématique animale. *Ann. Soc. R. Zool. Belg.* **83**, 111–30.

Foster, R. J. and G. M. Philip 1978. Tertiary holasteroid echinoids from Australia and New Zealand. *Palaeont.* **21**, 791–822.

Fricke, H. W. 1971. Fische als Feinde tropischer Seeigel. *Mar. Biol.* **9**, 328–38.

Gale, A. S. and A. B. Smith 1982. The palaeobiology of the Cretaceous irregular echinoids *Infulaster* and *Hagenowia. Palaeont.* **25**, 11–42.

Gezelius, G. 1963. Adaptation of the sea urchin *Psammechinus miliaris* to different salinities. *Zool. Bid. Uppsala* **35**, 329–37.

Ghiold, J. 1979. Spine morphology and its significance in feeding and burrowing in the sand dollar *Mellita quinquiesperforata* (Echinodermata: Echinoidea). *Bull. Mar. Sci.* **29**, 481–90.

Giese, A. C. and A. Farmanfarmaian 1963. Resistance of the purple sea-urchin to osmotic stress. *Biol Bull.* **124**, 182–92.

Giese, A. C., A. Farmanfarmaian, S. Hilden and P. Doezema 1966. Respiration during the reproductive cycle in the sea urchin *Strongylocentrotus purpuratus*. *Biol Bull.* **130**, 192–201.

Gladfelter, W. B. 1978. General ecology of the cassiduloid urchin *Cassidulus caribbearum*. *Mar. Biol.* **47**, 149–60.

Glynn, P. W. 1968. Mass mortalities of echinoids and other reef flat organisms coincident with midday low water exposures in Puerto Rico. *Mar. Biol.* **1**, 226–43.

Gordon, I. 1926. The development of the calcareous test of *Echinus miliaris*. *Phil Trans R. Soc. Lond. (B)* **214**, 259–312.

Gordon, I. 1929. Skeletal development in *Arbacia*, *Echinarachnius* and *Leptasterias*. *Phil Trans R. Soc. Lond. (B)* **217**, 289–326.

Hall, K. R. L. and G. B. Schaller 1964. Tool-using behaviour of the California sea otter. *J. Mammal.* **45**, 287–98.

Hancock, J. M. and E. G. Kauffman 1979. The great transgressions of the late Cretaceous. *J. Geol Soc. Lond.* **136**, 175–86.

Haude, R. and F. Langenstrassen 1976. *Rotasaccus dentifer* n.g. n. sp., ein devonischer Ophiocistioide (Echinodermata) mit 'holothuroiden' Wandskleriten und 'echinoidem' Kauapparat. *Paläont. Z.* **50**, 130–50.

Heatfield, B. M. 1965. Substrate preferences of the sea urchins *Arbacia punctulata* and *Strongylocentrotus droebachiensis*. *Biol Bull.* **129**, 407.

Hendler, G. 1977. The differential effects of seasonal stress and predation on the stability of reef-flat echinoid populations. *Proc. 3rd Int. Coral Reef Symp., Miami*, 217–223.

Hendler, G. 1978. Development of *Amphioplus abditus* (Verrill) (Echinodermata: Ophiuroidea). II, Description and discussion of ophiuroid skeletal ontogeny and homologies. *Biol Bull.* **154**, 79–95.

Hess, H. 1973. Neue Echinodermenfunde aus dem mittleren Dogger des Aargauer Juras. *Eclog. Geol. Helv.* **66**, 625–56.

Higgins, R. C. 1974. Observations on the biology of *Apatopygus recens* (Echinoidea: Cassiduloida) around New Zealand. *J. Zool.* **173**, 505–16.

Higgins, R. C. 1975. Variation in the number of anterior ambulacral tube-feet in *Echinocardium cordatum* (Pennant) (Echinoidea: Spatangoida) from Tasman Bay and Whangateau Harbour. *Recs NZ Oceanogr. Inst.* **2**, 141–7.

Higgins, R. C. 1976. Observations on the morphology of *Echinocardium cordatum* (Echinoidea: Spatangoida) from diverse geographical areas. *J. Zool.* **177**, 507–16.

Himmelman, J. H. 1978. Reproductive cycle of the green sea urchin *Strongylocentrotus droebachiensis*. *Can. J. Zool.* **56**, 1828–36.

Himmelman, J. H. and D. H. Steele 1971. Foods and predators of the green sea urchin *Strongylocentrotus droebachiensis* in Newfoundland waters. *Mar. Biol.* **9**, 315–22.

Hotchkiss, F. H. C. 1979. Studies on echinoderm ray homologies: Lovén's Law applies to Paleozoic ophiuroids. *J. Paleont.* **52**, 537–44.

Hyman, L. H. 1955. *The invertebrates*, vol. 4: *Echinodermata*. New York: McGraw-Hill.

Jackson, R. T. 1912. Phylogeny of the echini, with a revision of the Palaeozoic species. *Mem. Boston Soc. Nat. Hist.* **7**, 1–490.

Jangoux, M. and J. M. Lawrence (eds) 1982. *Echinoderm nutrition*. Rotterdam: Balkema.

Jensen, M. 1981. Morphology and classification of Euechinoidea Bronn, 1860 – a cladistic analysis. *Vid. Meddr Dansk Naturh. Foren.* **143**, 7–99.

Jesionek-Szymanska, W. 1963. Echinides irreguliers du Dogger de Pologne. *Acta Pal. Polon.* **8**, 293–414.

Jesionek-Szymanska, W. 1968. Irregular echinoids – an insufficiently known group. *Lethaia* **1**, 50–65.

Jesionek-Szymanska, W. 1979. Morphology and microstructure of oligolamellar teeth in Paleozoic echinoids. Part 1: teeth of some early lepidocentrid echinoids. *Acta Pal. Polon.* **24**, 265–74.

Kawamura, K. 1973. Fishery biological studies on the sea urchin *Strongylocentrotus intermedius* (Agassiz) (in Japanese with English abstract). *Sci. Rep. Hokkaido Fish. Exp. Statn* **16**, 1–54.

Kermack, K. A. 1954. A biometrical study of *Micraster coranguinum* and *M. (Isomicraster) senonensis. Phil. Trans. R. Soc. Lond. (B)* **237**, 375–428.

Khamala, C. P. M. 1971. Ecology of *Echinometra mathaei* (Echinodermata; Echinoidea) at Diani Beach, Kenya. *Mar. Biol.* **11**, 167–172.

Kier, P. M. 1956. Separation of interambulacral columns from the apical system in the Echinoidea. *J. Paleont.* **30**, 971–4.

Kier, P. M. 1962. Revision of the cassiduloid echinoids. *Smithson. Misc. Colls* **144**, 1–262.

Kier, P. M. 1965. Evolutionary trends in Paleozoic echinoids. *J. Paleont.* **39**, 436–65.

Kier, P. M. 1968. Echinoids from the Middle Eocene Lake City Formation of Georgia. *Smithson. Misc. Colls* **153**, 1–45.

Kier, P. M. 1969. Sexual dimorphism in fossil echinoids. In *Sexual dimorphism in fossil metazoa and taxonomic implications*. G. E. G. Westermann (ed.), 215–222. Stuttgart: Schweizerbart'sche.

Kier, P. M. 1970. Lantern support structures in the clypeasteroid echinoids. *J. Paleont.* **44**, 98–109.

Kier, P. M. 1974. Evolutionary trends and their functional significance in the post-Paleozoic echinoids. *J. Paleont.* **48** (suppl.): *Paleont. Soc. mem.* **5**, 1–95.

Kier, P. M. 1977. The poor fossil record of the regular echinoid. *Paleobiol.* **3**, 168–74.

Kier, P. M. 1981. A bored Cretaceous echinoid. *J. Paleont.* **55**, 656–9.

Kier, P. M. 1982. Rapid evolution in echinoids. *Palaeont.* **25**, 1–10.

Kier, P. M. and R. E. Grant 1965. Echinoid distribution and habits, Key Largo coral reef preserve, Florida. *Smithson. Misc. Colls* **149**, 1–68.

Kier, P. M. and M. H. Lawson 1978. Index to living and fossil echinoids, 1924–1970. *Smithson. Contrib. Paleobiol.* **34**, 1–182.

Kryuchkova, G. A. and A. N. Solov'yev 1975. The larval stages of echinoids. *Paleont. J.* **9**, 487–93.

Lambert, J. 1933. Échindes de Madagascar communiques par M. H. Besaire. *Annls Géol. Serv. Mines, Madagasc.* **3**, 1–49.

Lambert, J. and P. Thiéry 1909–25. *Essai de nomenclature raisonnée des échinides.* Chaumont: Ferrière.

Lane, J. E. M. and J. M. Lawrence 1980. Seasonal variation in body growth, density and distribution of a population of sand dollars, *Mellita quinquiesperforata* (Leske). *Bull. Mar. Sci.* **30**, 871–82.

Lawrence, J. M. 1975. On the relationships between marine plants and sea urchins (Echinodermata; Echinoidea). *Oceanogr. Mar. Biol. Ann. Rev.* **13**, 213–87.

Lawrence, J. M. and I. Ferber 1971. Substrate particle size and the occurrence of *Lovenia elongata* (Echinodermata: Echinoidea) at Taba, Gulf of Elat (Red Sea). *Israel J. Zool.* **20**, 131–8.

Lewis, D. N. and P. C. Ensom 1982. *Archaeocidaris whatleyensis* sp. nov. (Echinoidea) from the Carboniferous Limestone of Somerset and notes on echinoid phylogeny. *Bull. Br. Mus. Nat. Hist. (Geol.)* **36**, 77–104.

MacBride, E. W. 1906. Echinodermata. In *The Cambridge natural history*, S. F. Harmer and A. E. Shipley (eds). London: Macmillan.

MacBride, E. W. and W. Spencer 1938. Two new Echinoidea, *Aulechinus* and *Ectinechinus*, and an adult plated holothurian *Eothuria* from the Upper Ordovician of Girvan. *Phil. Trans. R. Soc. Lond. (B)* **229**, 91–136.

Mannil, R. 1962. The taxonomy and morphology of *Bothriocidaris* (Echinoidea) (in Russian with an English abstract). *Ensv. Teaduste Akad. Geol. Inst. Uurim* **9**, 143–90.

Märkel, K. 1979. Structure and growth of the cidaroid socket joint lantern of Aristotle compared to the hinge-joint lanterns of non-cidaroid regular echinoids (Echinodermata; Echinoidea). *Zoomorph.* **94**, 1–32.

Märkel, K. 1981. Experimental morphology of coronal growth in regular echinoids. *Zoomorph.* **97**, 31–52.

Märkel, K., P. Gorny and K. Abraham 1977. Microarchitecture of sea urchin teeth. *Fortsch. Zool.* **24**, 103–114.

Märkel, K., F. Kubanek and A. Willgallis 1971. Polykirstalliner Calcit bei Seeigeln (Echinodermata: Echinoidea). *Z. Zellforsch.* **119**, 355–77.

Matsumura, T., M. Hasegawa and M. Shigei 1979. Collagen biochemistry and phylogeny of echinoderms. *Comp. Biochem. Physiol.* **62B**, 101–5.

Mayr, E. 1954. Geographic speciation in tropical echinoids. *Evolution* **8**, 1–18.

McCoy, F. 1849. On some new Palaeozoic Echinodermata. *Ann. Mag. Nat. Hist. Ser. 2* **3**, 244–54.

McNamara, K. J. and G. M. Philip 1980. Australian Tertiary schizasterid echinoids. *Alcheringa* **4**, 47–65.

McNulty, J. K., R. C. Work and H. B. Moore 1962. Level sea bottom communities in Biscayne Bay and neighbouring areas. *Bull. Mar. Sci.* **12**, 204–33.

Michalik, J. 1977. Systematics and ecology of *Zeilleria bayle* and other brachiopods in the uppermost Triassic of the west Carpathians. *Geol. Zbornik–Geol. Carpath.* **28**, 323–46.

Mintz, L. W. 1968. Echinoids of the Mesozoic families Collyritidae d'Orbigny and Disasteridae Gras. *J. Paleont.* **42**, 1272–88.

Mironov, A. N. 1980. Two modes of formation of deep-sea echinoid fauna. *Oceanol.* **20**, 462–5.

Moore, D. R. 1956. Observations on predation on echinoderms by three species of Cassidae. *Nautilus* **69**, 73–6.

Moore, H. B. 1966. Ecology of echinoids. In *Physiology of Echinodermata*, R. A. Boolootian (ed.), 73–86. London: Wiley Interscience.

Mudge, D. C. 1978. Stratigraphy and sedimentation of the Lower Inferior Oolite of the Cotswolds. *J. Geol Soc. Lond.* **135**, 611–27.

Müller, A. H. 1970. Über den Sexualdimorphismus regulärer Echinoiden (Echinodermata) der Oberkreide. *Mber. Dt. Akad. Wiss. Berlin* **12**, 923–35.

Nichols, D. 1959a. Changes in the chalk heart-urchin *Micraster* interpreted in relation to living forms. *Phil. Trans. R. Soc. Lond. (B)* **242**, 347–437.

Nichols, D. 1959b. Mode of life and taxonomy in irregular sea-urchins. *Spec. Publ. Syst. Assoc.* **3**, 61–80.

Nichols, D. 1972. The water vascular system in living and fossil echinoderms. *Palaeont.* **15**, 519–38.

O'Neill, P. L. 1978. Hydrodynamic analysis of feeding in sand dollars. *Oecologia* **34**, 157–74.

Palmer, T. J. 1982. Cambrian to Cretaceous changes in hardground communities. *Lethaia* **15**, 309–324.

Parks, N. B. 1973. Distribution and abundance of the sand dollar *Dendraster excentricus* off the coast of Oregon and Washington. *Fish. Bull. Natn. Oceanic Atmos. Admin. US* **71**, 1105–9.

Patterson, C. and D. E. Rosen 1977. Review of ichthyodectiform and other Mesozoic teleost fishes and the theory and practice of classifying fossils. *Bull. Am. Mus. Nat. Hist.* **158**, 81–172.

Paul, C. R. C. 1967. New Ordovician Bothriocidaridae from Girvan and a reinterpretation of *Bothriocidaris* Eichwald. *Palaeont.* **10**, 525–41.

Pearse, J. S. 1969. Reproductive periodicities of Indo-Pacific invertebrates in the Gulf of Suez. II, the echinoid *Echinometra mathaei* (de Blainville). *Bull. Mar. Sci.* **19**, 580–613.

Pearse, J. S. and V. B. Pearse 1975. Growth zones in echinoid skeleton. *Am. Zool.* **15**, 731–53.

Pearse, J. S. and B. F. Phillips 1968. Continuous reproduction in the Indo-Pacific sea urchin *Echinometra mathaei* at Rottnest Island, Western Australia. *Aust. J. Mar. Freshwat. Res.* **19**, 161–72.

Pequignat, E. 1970. Biologie des *Echinocardium cordatum* (Pennant) de la Baie de Seine. Nouvelles recherches sur la digestion et l'absorption cutanées chez les échinides et les stellerides. *Forma et Functio* **2**, 121–68.

Philip, G. M. 1963. Silurian echinoid pedicellariae from New South Wales. *Nature* **200**, 1334.

Philip, G. M. 1965. Classification of echinoids. *J. Paleont.* **39**, 45–62.

Philip, G. M. and R. J. Foster 1971. Marsupiate Tertiary echinoids from south-eastern Australia and their zoogeographic significance. *Palaeont.* **14**, 666–95.

Randall, J. E. 1967. Food habits of reef fishes of the West Indies. *Stud. Trop. Oceanogr. Univ. Miami* **5**, 665–847.

Raup, D. M. 1956. *Dendraster*: a problem in echinoid taxonomy. *J. Paleont.* **30**, 685–94.

Raup, D. M. 1962. The phylogeny of calcite crystallography in echinoids. *J. Paleont.* **36**, 793–810.

Rose, E. P. F. 1978. Some observations on the Recent holectypoid echinoid *Echinoneus cyclostomus* and their palaeoecological significance. *Thalass. Jugoslav.* **12**, 299–306.

Rowe, A. W. 1899. An analysis of the genus *Micraster*, as determined by rigid zonal collecting from the zone of *Rhynchonella cuvieri* to that of *Micraster coranguinum*. *Q. J. Geol Soc. Lond.* **55**, 494–547.

Seilacher, A. 1979. Constructional morphology of sand dollars. *Paleobiol.* **5**, 191–221.

Schäfer, W. 1971. *Ecology and palaeoecology of marine environments*. Edinburgh: Oliver and Boyd.

Shepherd, S. A. 1973. Competition between sea urchins and abalones. *Austral. Fish.* **32**, 4–7.

Smith, A. B. 1978a. A comparative study on the life-styles of two Jurassic irregular echinoids. *Lethaia* **11**, 57–66.

Smith, A. B. 1978b. A functional classification of the coronal pores of regular echinoids. *Palaeont.* **21**, 759–90.

Smith, A. B. 1979. Peristomial tube feet and plates of regular echinoids. *Zoomorph.* **94**, 67–80.

Smith, A. B. 1980a. The structure and arrangement of echinoid tubercles. *Phil. Trans. R. Soc. Lond. (B)* **289**, 1–54.

Smith, A. B. 1980b. The structure, function and evolution of tube feet and ambulacral pores in irregular echinoids. *Palaeont.* **23**, 39–84.

Smith, A. B. 1980c. Stereom microstructure of the echinoid test. *Spec. Pap. Palaeont.* **25**, 1–81.

Smith, A. B. 1981. Implications of lantern morphology for the phylogeny of post-Palaeozoic echinoids. *Palaeont.* **24**, 779–801.

Smith, A. B. and T. P. Crimes 1983. Trace fossils formed by heart-urchins (Echinoidea): a study of *Scolicia* and related traces. *Lethaia* **16**, 79–92.

Smith, A. G., J. C. Briden and G. E. Drewry 1973. Phanerozoic world maps. *Spec. Pap. Palaeont.* **12**, 1–42.

Spencer, W. K. 1914–40. A monograph of the British Palaeozoic Asterozoa. *Palaeontogr. Soc. (Monogr.)*.

Stanton, R. J., J. R. Dodd and R. R. Alexander 1979. Excentricity in the clypeasteroid echinoid *Dendraster*: environmental significance and application in Pliocene paleoecology. *Lethaia* **12**, 75–87.

Steen, J. B. 1965. Comparative aspects of the respiratory gas exchange of sea urchins. *Acta Physiol. Scand.* **63**, 164–70.

Stokes, R. B. 1975. Royaumes et provinces faunistique de Crétace établis sur la base d'une étude systématique du genre *Micraster*. *Mém. Mus. Natn. Hist. Nat. (C)* **31**, 1–94.

Stokes, R. B. 1977. The echinoids *Micraster* and *Epiaster* from the Turonian and Santonian chalk of England. *Palaeont.* **20**, 805–21.

Stokes, R. B. 1979. An analysis of the ranges of spatangoid echinoid genera and their bearing on the Cretaceous/Tertiary boundary. In *Cretaceous–Tertiary boundary events*, W. K. Christensen and T. Kirkelund (eds), 78–82. Copenhagen: University of Copenhagen.

Strathmann, R. R. 1981. The role of spines in preventing structural damage to echinoid tests. *Paleobiol.* **7**, 400–6.

Thompson, R. J. 1983. The relationship between food ration and reproductive effort in the green sea urchin *Strongylocentrotus droebachiensis*. *Oecologia* **56**, 50–7.

Timko, P. L. 1976. Sand dollars as suspension feeders: a new description of feeding in *Dendraster excentricus*. *Biol Bull.* **151**, 247–59.

Weber, J. N. 1969. The incorporation of magnesium into the skeletal calcites of echinoderms. *Am. J. Sci.* **267**, 537–66.

Weber, J. N. 1973. Temperature dependence of magnesium in echinoid and asteroid skeletal calcite: a reinterpretation of its significance. *J. Geol.* **81**, 543–56.

Weber, J. N., R. Greer, B. Voight, E. White and R. Roy. 1969. Unusual strength properties of echinoderm calcite. *J. Ultrastruct. Res.* **26**, 355–66.

Weihe, S. C. and I. E. Gray 1968. Observations on the biology of the sand dollar *Mellita quinquiesperforata* (Leske). *J. Elisha Mitchell Sci. Soc.* **84**, 315–27.

Zinsmeister, W. J. 1980. Observations on the predation of the clypeasteroid echinoid, *Monophoraster darwini*, from the Upper Miocene Entrerrios Formation, Patagonia, Argentina *J. Paleont.* **54**, 910–2.

Zoeke, E. 1951. Étude des plaques des *Hemiaster*. *Bull. Mus. Natn. Hist. Nat.* **23**, 696–705.

Index

Numbers in **bold** type refer to text sections, and numbers in *italic* type refer to text figures.

Irregularia

Atelostomata

Echinacea

...steroida — **Holasteroida** — **Spatangoida** — **Stirodonta** — **Camarodonta**

Protoscutellidae
Echinarachnidae
Scutellidae
Astriclypeidae
Monophorasteridae
Mellitidae
Dendrasteridae

Urechinidae
Calymnidae
Pourtalesidae
Stenonasteridae
Holasteridae
Somaliasteridae
Micrasteridae
Toxasteridae
Brissidae
Spatangidae
Loveniidae
Hemiasteridae
Pericosmidae
Palaeostomatidae
Aeropsidae
Schizasteridae
Asterostomatidae
?

Saleniidae
Acrosaleniidae
Hemicidaridae
Phymosomatidae
Stomechinidae
Arbaciidae
Pseudodiadematidae
Glyphocyphidae

Temnopleuridae
Toxopneustidae
Echinometridae
Echinidae
?
Strongylocentrotidae
Paraechinidae
?

Collyritidae
Disasteridae

?

...anges of echinoid families.